New Paths to God and the Soul

Roger Skrenes

En Route Books and Media, LLC
St. Louis, MO

ENROUTE
Make the time

En Route Books and Media, LLC
5705 Rhodes Avenue
St. Louis, MO 63109

Cover credit: TJ Burdick Designer

Library of Congress Control
Number: 2020930380

ISBN-13: 978-1-950108-93-0

Table of Contents

Part Two: Soul in the Bible

Part Three: Soul and Knowledge

Part Four: Related Topics

Introduction

The human body consists of many material parts: cells, tissues, and organs. The brain itself consists of a collection of functioning parts. A person, however, is certain that he or she is not a collection of parts, but is rather one whole entity, an individual. A person is not just a body, consisting of various parts. In fact, most people feel that they are in some way distinct from their own body. The body is *used* by them but cannot be said to *be* them. At certain times a person may even feel like a prisoner in his or her own body, especially during times of illness. No one should, therefore, make the mistake of thinking a person is solely their body. He or she is something more. That something more is the soul.

The real thinking and acting person is better thought of as a soul that forms a body even though the soul is not visible! Researchers do not know how to view the soul because it cannot be detected by the senses. One cannot see it or touch it or determine its shape. It is like trying to describe or determine the shape or size of a thought or a wish. Thus, the soul is spoken of as spirit-like, being immaterial because it has no shape or size and no parts. And because it has no parts, it cannot die! When the body dies, the physical parts of the body begin to separate and eventually dissolve. However, since the soul has no parts, it cannot die. It is, in fact, the "spirit," or "form," of the body, so that when the soul leaves the body, the body dies.

The *Catechism of the Catholic Church* (1994) has the following words to say about the nature and origin of each person's soul.

> #365
> The unity of soul and body is so profound that one has to consider the soul to be the 'form' of the body: that is, it is because of its spiritual soul that the body ... becomes a living, human body ... not two natures united, but rather their union forms a single [human] nature.
>
> #366
> The church teaches that every spiritual soul is created immediately [at conception] by God – it is not 'produced' by the parents – and also that it [the soul] is

> immortal. It does not perish when it separates from the body at death, and it will be reunited with the body at the general resurrection.

It is God, therefore, who gives each human body a unique form, or soul.

> Yahweh God [the Father] formed [the body of] man of dust from the ground, and breathed into his nostrils the breath of life [his soul].
>
> Genesis 2:7

This book has two objectives. First, it seeks to revisit the question of God's existence, adding some insights from twentieth century science. Second, it will explore various pieces of evidence for the existence of the human soul. It will examine the operations of the intellect and will of the individual person. It will discuss dreams, imagination, and the mysteries of matter as it mimics the soul. Furthermore, it will examine memory and the various kinds of bodies (from infancy to old age) that we use and discard as we live upon this earth. Lastly, it will examine "out of body experiences" and certain discoveries made by neuroscience that involve the soul.

This two-fold plan was chosen because if

God exists then both his power and intent will be present to ensure individual life beyond death. If God is eternal and we are his handiwork, then we will resemble him in this way. That is what is revealed in the Bible.

> God created man in His image. In the image of God He created him.
>
> Genesis 1:27

Because God is the Living One, it is logical that a creature created in His image will live beyond death – will, like God, continue to live. Hence, the existence of the soul is affirmed by the very work and promises of God.

Part One:

God's Existence

1.

First Cause and the Big Bang

In 1917, the world's largest telescope, having a 100-inch reflecting mirror, began operations on Mt. Wilson outside of Los Angeles, California. Edwin Hubble came to this location two years later, in 1919, to study the hundreds of "nebulae" seen in the night sky. These nebulae were cloud-like patches within various constellations of stars that did not seem to move. No one knew what they were. However, by 1923 Hubble was able to isolate certain individual stars within one such area of the night sky, the "Andromeda nebula." Hubble then set himself the task of determining the distance to the stars in the "Andromeda Nebula." To do this, he used a group of very bright stars known as "Cepheid Variables." These stars had specific peaks of brightness, and the distances to a few of them were known within our own Milky Way galaxy.

Night after night, he took photographs of Andromeda with the enormous telescope, searching for Cepheids. In October, 1923, he found one such star blinking in one of Andromeda's spiral arms. The light from this star made it clear that Andromeda was well outside the bounds of the Milky Way.

Hubble also found other Cepheids present within the Andromeda nebula. So, by comparing the brightness of Cepheid stars within our own galaxy to their brightness in the Andromeda nebula, Hubble was able to estimate that Andromeda's distance was about 900,000 light years away from us.

On Dec. 30, 1924, Hubble announced that the spiral nebula Andromeda was actually a galaxy and that our own Milky Way galaxy is just one of many galaxies in the universe. Hubble had found that many of the nebulae he studied were really galaxies like the Andromeda.

In 1926, Hubble began an investigation of the electromagnetic signature of these various galaxies. By isolating their "red shifted" light, he was able to determine the speed of the various galaxies. And by 1929 he had announced that all the galaxies he had studied were rapidly speeding away from one another. In other words, the universe, with its many galaxies, was "expanding" in all directions away from a single point.

The situation resembled a balloon with dots on it. The dots represent galaxies expanding as the balloon is inflated. As the expansion progresses, the galaxies are seen to be going faster and faster away from each other. Then, when the air is let out of the balloon, it decreases in size down to a small point. The huge explosion that is believed to have happened when the universe began has come, thanks to the work of Fr. Georges Lemaître, SJ, to be known to the world as the "Big Bang."

This "Big Bang" model of creation mirrored in some ways an earlier rational structure sketched out by St. Thomas Aquinas about 1260 AD. Aquinas spoke of God as the "Uncaused Causer" and the "Unmoved Mover" of the world. Aquinas's two pathways coincide in part with the "Big Bang" origin of the universe.

Aquinas's "Uncaused Causer" reasoning looks somewhat like the following.

1. Humans see many causes of things in the world.
2. Nothing however, is the cause of its own existence. If this were so, it would have to exist prior to itself, to make itself come into existence. This, of course, is impossible.
3. Aquinas reasoned further that if the first thing in a long series of causes did not exist, nothing in that series could exist. The principal is this: "from nothing comes nothing."
4. If the series of secondary causes extended infinitely into the past without a First Cause, there would also be nothing exist-

ing today – again because "nothing will cause nothing to exist."

5. But there are many things existing
6. Therefore, there was a First Cause at the origin of the universe, who was Himself uncaused. That First Cause of everything is God.

One question that sometimes arises in this kind of discussion is: If God is the fundamental cause of everything in the universe, then who or what caused God to exist? The answer to this question is as follows. First, God did not start existing! God has *always* existed. God is eternal. Second, God is not one being among many in the universe. God is Being itself! For example, God said of himself in the burning bush, when speaking to Moses: "I Am Who Am!" (Exodus 3:14) In other words, God is the existent One. Or to put it another way: God is Being itself, the source of all being. He is outside the world of ordinary being, which he has created and continues to create. Lastly, God, from minute to minute, keeps all being in existence!

The second pathway that Aquinas spoke of mirrors the motion of the "expanding universe." This pathway involved an "Unmoved Mover." Here is an approximation of Aquinas's thought:

1. Our senses tell us that things in this world are in motion. Motion, however, means more than simply moving from place to place. A thing is said to be in motion if it is acquiring something it did not formerly have. For example, when a plant is growing it is moving; when a student acquires knowledge he or she is moving; and when a person is advancing in age he or she is also moving.
2. Things move when a potential (or possible) motion becomes an actual (or happening) motion.
3. Only something moving can put something not moving into motion.
4. Nothing can move itself without help. Each thing in motion is moved initially by something else.
5. This sequence of causal motion, however, cannot go on forever.
6. There is at some point a First Mover, whose own motion is caused by nothing (or by no one) else.
7. This powerful "Unmoved Mover" is God.

2.

Omniscient Designer

Intelligent beings in the world consciously set goals for themselves, while non-intelligent beings do not. Non-intelligent beings, however, live as if they were guided by some great unseen intelligence. They appear to accomplish goals that they cannot and do not set for themselves. This is especially true of non-living things, like the chemical elements, which follow precise chemical rules. This then is the heart of St. Thomas Aquinas's reasoning concerning God as the Great Designer:

1. Most things in nature do not have knowledge. In most cases, they do not even have brains.
2. However, things in nature generally work toward some goal, and they do not do this by mere chance.
3. If things in nature operated by chance, they would not have predictable results.

4. However, much of the behavior in nature is set, and precise. Yet it cannot be set by themselves since they are non-intelligent. (An illustration of this might be that of an arrow which is non-intelligent but reaches its target because of an intelligent archer.)
5. Therefore, this orderly behavior in nature must be set by something else (for example, secondary causes) or by someone else (the primary cause).
6. This primary cause, or infinitely intelligent Being, who guides all non-intelligent things in nature to their goal, is God.

3.

Necessary Existent

In his "way to God," St. Thomas Aquinas addresses the topic of "contingency" in our world. Contingency means that all things, including all living things, exist on the earth only temporarily. They come into existence and then go out of existence. This includes every living thing that evolution speaks of. Animals die, men die, even buildings erected by humans will eventually come down. The world as we see it around us is, in fact, passing away. Everything within this world is contingent.

Therefore, at its beginning, how did our world get started? And how does it continue in existence? Aquinas begins by saying that everything in our world has a "form." This structural beginning is contained in the mind of God. It is the "potential" start of all things and beings in our world.

Existence, however, is a condition or state that is separate from the form of a particular thing or being. Things can exist in the mind of God that do not necessarily exist materially on earth. For example, unicorns do not exist in our world and never did. Dinosaurs, however, existed earlier in our world, but no longer do exist.

Existence itself is a power possessed only by God. Within God himself, "essence" is joined to "existence." In other words, God's essence and existence are one. God is the Necessary Existent. God's essence is existence itself. He is the Existent One who gives existence to all other things and beings in our world. To say this another way: all temporal things and beings in our world get their continuing existence from God. Their forms become "actualized" in their existence, and their ongoing presence in the world is due to the power of God. There are secondary causes of existence in our world for sure, but the fundamental origin and continuance of all existence in our universe is from God. He is the one whose essence is existence. He is the Necessary Existent.

Through the Word of Yahweh
[God the Father] the heavens were made;
and by the breath of his mouth all their host.

Psalm 33:6

In the beginning was the Word ...
and God [the Son. Jesus] was the Word ...
All [things] became through Him ...

John 1:1-3

God ... calling the not being, as being ...

Romans 4:17

In him we live and are moved and have are being ...

Acts 17:28

4.

Intelligible Universe

The intelligibility of the world is persuasive evidence that a great intelligence, namely God, has thought the world into being. Such a way of seeing God's presence envisions a link between the sciences and religion. Every scientist, whether a physicist, chemist, biologist, or psychologist, assumes that the world around them is intelligible. This means that the world that is seen or unseen can be known. Each scientist goes out to meet a world that is imbued or permeated with meaning. Such intelligibility is present throughout the cosmos.

The best explanation for why the universe is meaningful is that it has been thought, and then spoken, into existence by God. God is not in the universe he has created, but he is the power and intelligence behind its formation. He is Being itself, the very act of Being, that brings all things into existence from nothing. The world is not just "dumbly" present; the world is rather filled with reason and mind. That is why a person can "re-cog-nize" (recognize) truth when he or she sees it. The truth about a thing is already there, waiting to be discovered by the person doing research. Therefore, in some sense, the researcher participates in the mental workings of God himself.

5.

Mathematics

Certain truths within the universe are both necessary and eternal. A good example of this is found in mathematics. Items like squares, triangles, and circles are not dependent on the human mind for their existence. Shapes such as these have been discovered rather than invented by men.

Mathematical truths in general are unalterable and necessary. For example, the operation 2 + 2 = 4 was true before humans arrived on this planet and will remain true even if all humans were to die out. The fact that the angles within a triangle add up to 180 degrees has always been true. It is a necessary truth that is also eternal. Math objects and laws are, in fact, eternal truths that cannot have started to be true in the past and then perhaps at some time in the future will cease to be true. They are necessary truths, eternal truths.

St. Augustine understood this to mean that such eternal truths, represented especially within mathematics, could only have originated in an eternal Mind. Such eternal truths certainly did not come into existence from nothing! Augustine said further that only God has such a Eternal Mind. In other words, eternal truths require the existence of the Eternal Mind of God.

> Mathematics is the language in which God has written the universe.
>
> Galileo c.1623

6.

Truth and Goodness

There are natural desires and man-made desires. Man-made desires are not necessary; for example, buying an expensive shirt or dress.

Natural desires are, however, ever present; our need for food and water are examples. The fact that we have a real need for food points to the certainty that food will be found in the environment. And, indeed, we do find many plants that are good sources of food. Likewise, our need for water points to a certainty that it will be found in rainfall, or in nearby streams and lakes. These necessary desires make possible one's life here on earth.

We also have necessary desires that point to life in Heaven with God. Two of these supernatural desires, that are on display within us, are our desire to know truth and our desire for goodness. The desire to know the truth of things begins early in life. Every parent experiences being asked many questions by their small children. Answers are given and more questions are asked. And this pattern continues throughout life. Later, when the scientist discovers something new, several additional questions will be generated that require his or her attention. Thus, the search for the ultimate truth goes on, and on, and on. That is so because the ultimate truth is God, who is not fully revealed on this side of life.

The desire for good within our personal life, and for the world around us, is also a fact of life. We seek the good for ourselves and a just set of circumstances for those around us whom we love. This sense of what is good and just also begins early in life and continues forward. The child will say, for example, to his or her parents: "That's not fair!" This will occur even before the child goes to school! A child seems to know what "good" or "fair" means without being taught! And such a movement toward the good, or the fair, goes on throughout one's life. Nonetheless, its ultimate end or resolution cannot be obtained here on earth – even though it is attempted. Like truth, the ultimate good is God himself, partially known in this world by his "footprints" — his deeds and creations. In other words, the complete good is what we seek with our small efforts toward goodness, and this complete good is God.

7.

Ascent to Perfection

In this world there are differing levels of perfection. For example, there are different levels of truth, different levels of love, different levels of goodness, and different levels of beauty. These differing levels have an ascendency pattern: from little, to more, to much, and then to most. One example: Consider people's knowledge of the hydrogen atom. The elementary student knows very little about the atom, the high school student knows more, the college student has a considerable knowledge, and the Doctor of Physics knows the most. However, the more we learn about subatomic particles, the more questions we have! This ascent continues onward into the unknown and ultimately to God its creator. Thus, the path of knowledge goes from very little knowledge of truth to ultimate truth.

Such is the case not only with truth, but also with love, goodness, and beauty. Each of these topics points upwards toward God and is one of the signs of his existence.

Love and goodness seem to go hand in hand. First, there is love of the self only. Next there is love of the self in competition with others. Then there is love of the self to help others. Next, there is love of others, with a minimum of concern about oneself. And lastly, there is mostly a concern about others together with the love of God.

In the area of beauty, adjectives exist to indicate its various levels. First, there is the "cute," then there is the "pretty," followed by the "attractive," and finally the "beautiful." God encompasses all of these categories, but especially the beautiful. The visual arts and music often call to mind God's divine Presence.

St. Thomas Aquinas said that the "beautiful" possess three basic characteristics: wholeness, harmony and radiance. What follows are a few lines in the Bible that speak of the upward ascent to God – in truth, in love, in goodness, and in beauty.

Truth

"He who walks uprightly ... speaks the truth ..."

Psalm 15:2

"Those who do truth are His delight."

Proverbs 12:22

"The Word [of God] became flesh ... full of ... truth."

John 1:14

"Jesus answered: 'I am ... the truth ...'"

John 14:6

Love

"Love does no harm to the neighbor."

Romans 13:10

"Love is patient, love is kind. It does not envy. It does not boast. It is not proud. It does not dishonor others ... It keeps no record of wrongs. Love does not delight in evil but rejoices with the truth."

1 Cor. 13:4-6

"Let us not love with words ... but with actions ..."

1 John 1:6

"Through love, serve one another ..."

Galatians 5:13

"Love one another, because love is from God ..."

1 John 4:7

"Keep yourself in the love of God, awaiting ... eternal life."

Jude 21

"God is love."

1 John 4:8

Goodness

"We are God's handiwork ... to do good works."

Ephesians 2:10

"Hate what is evil, hold fast to what is good."

Romans 12:9

"Give thanks to the Lord, for He is good."

Psalm 107:1

"There is only One who is [completely] good [God]."

Matthew 19:1

Beauty

"He [God] made everything beautiful in its time."

Eccles. 3:11

"She is more precious than jewels."

Prov 3:15

"Solomon in all his splendor was not dressed like one of these [flowers]."

Matt 6:29

"The Lord is... a crown of glory and of beauty."

Isaiah 28:5

"I will seek my dwelling in the house of the Lord ... to behold the beauty of the Lord."

Psalm 27:4

8.

God Fine-Tunes the Universe

There are many evidences of God's creative presence and power in the material universe. Such is the understanding presented in the Bible.

> The unseen [presence] of God from the creation
> of the world is seen through the things He
> has made, by His eternal power and deity ...
>
> Romans 1:20

Some of the evidences of God's continuing presence can be seen by how exactly his creation is fine tuned. There are many very tight specifications in the design of the created world. In the chapters below, a few of these exact tolerances will be examined. Many of these examples look anything but accidental! The physicist, Fred Hoyle of England, celebrated for his discoveries of how the heavier chemical elements are formed within stars, stated the following.

> A commonsense interpretation of the facts [in nature] suggests that a super intellect [God] has monkeyed with the physics, as well as with the chemistry and biology ... there are no blind forces worth speaking about in nature.

What follows will be a few of these fine-tuned examples of which Hoyle speaks.

9.

Strong Nuclear Force

The strong nuclear force holds the center of every atom together. Inside each proton and each neutron of an atom's nucleus are three quark particles. These quark particles are in turn held together by massless gluon particles. Gluons make up the strong nuclear force. The strong force binds the nucleus as a whole together.

The fine-tuning of the strong nuclear force is especially important when one speaks of the protons inside the nucleus of all atoms. There are 92 basic elements — or different kinds of atoms — in nature. Atom #1 has one [+] proton charged particle and is the atom we call hydrogen. Atom #2 has two [+] charged proton particles and is the atom helium. Atom #3 has three [+] proton charged particles and is the atom Lithium. The "number" of each atom tells how many proton particles are in that atom, all the way up to element 92, which is the Uranium atom with 92 [+] charged protons.

Two atoms which are very important for humans

are carbon and oxygen. These two atoms make up roughly 75% of the weight of a human body. Carbon is atom #6, with six proton particles. Oxygen is atom #8 with eight proton particles.

The fine-tuning of proton particles within atoms is critical because they are all positively [+] charged. Because they are [+] charged, all proton particles tend to repel (or push away from) one another within the nucleus. Therefore, without any counteracting force, the nucleus would have a tendency to fly apart!

The strong nuclear force, however, makes these [+] charged proton particles stick together. To do this, the gluon force that holds the [+] charged proton particles together must be precisely balanced. If, for example, the strong nuclear force were only *nine per cent weaker*, protons would not stick together, with the result that only one element, hydrogen, would exist in the universe. (Hydrogen, atom #1, has only one proton in its nucleus.) Each of the other atoms in nature with more than one proton particle would not be able to form or hold together. The result would be that the heavier elements like carbon and oxygen would not exist. Therefore, humans would not exist!

If, however, the strong nuclear force were only *two per cent stronger* than it is now, protons would have such an attraction for each other that a single proton (such as the one in a hydrogen atom) could not exist alone. In other words, there would be no hydrogen atoms in the universe because only heavier elements with more than one proton could form!

Therefore, life would become impossible,

because hydrogen is an essential component of all life, for example, in water, in DNA, and in all protein formation within the human body.

Hydrogen is also the primary fuel within stars, allowing stars to shine. Thus, one can appreciate the overarching presence of the "super intellect" of God within creation, of which Fred Hoyle spoke.

10.

Electromagnetic Force

The protons in the nucleus of every atom in the universe have a positive [+] charge. This positive charge is balanced by the negative [-] charge of an equal number of electron particles flying around the nucleus of the atom. The number of proton [+] particles is equal to the number of electron [-] particles in all atoms. Therefore, the atom as a whole is electrically neutral.

However, if the situation were to change only slightly, atoms would not be able to share electrons with other atoms.

If, for example, the positive, attractive [+] force of the nucleus were somewhat *larger* than it presently is, atoms would tightly "hang on to" the electrons flying around their perimeter. Thus, there would be no sharing of electrons with other atoms to form molecules, and therefore there would be no chemical compounds to make "life" possible.

However, if the positive [+] force of the nucleus were only slightly *weaker*, atoms would not hold on to their electrons at all. Electrons would become detached from the atom's outer structure. Thus, the sharing of electrons among different kinds of atoms would not be possible, and thousands of different molecules that are necessary in any living organism would not be possible. In other words, life could not exist, and we would not be here.

One example of the importance of having different atoms share electrons is seen in the water molecule. In water, two hydrogen atoms share electrons with one oxygen atom, forming the H_2O molecule. Since the human body is about two-thirds water, this sharing of electrons in the water molecule is extremely important. Without water, plants and animals of all kinds would not be able to exist on this planet.

11.

Gravity in Stars

The force of gravity causes hydrogen gas and dust in space to pull together into balls and form stars. As the gas and dust become more and more compressed, the center of the star gets hotter and hotter. Then, when the core temperature reaches several million degrees, hydrogen nuclei (element #1) begin to fuse together to form the heavier helium nuclei (element #2). In this process, about 0.7% of the mass of the hydrogen material is turned into heat and light energy, according to Einstein's equation $E=MC^2$. The energy (E) released by the fusion causes the star to begin to shine and give off large amounts of heat and light. The outward pressure of this fusion energy (evidenced by the heat and light) then balances the inward force of gravity, which allows the star to begin a long stable life.

The fine-tuning of stars essentially concerns gravity. The strength of the inward pull of gravity initially determines how hot the nuclear furnace at the core of a star will become. If the force of gravity were *too strong*, then the core of the star would be so hot that its hydrogen fuel would fuse at a very rapid pace and release energy too quickly. Such a star would be short lived, having only a million or a few million years of life. The result of such a short stellar life would be that all heavier elements such as carbon, oxygen, and iron would not have enough time to form. Also, a short-lived star would not allow the planets that circle around it enough time to develop complex forms of life, such as human beings. Stars like our sun, which is presently five billion years old, do in fact allow enough time for complex life forms to come into existence.

On the other side of this story are the would-be stars that have *too little* gravity to properly form. If the inward gravitational force that causes hydrogen gas and dust to gather into a ball were weaker than required, stars would never become hot enough inside to ignite or begin nuclear fusion. Thus, the collapsing mass would never become a star and would never give off heat or light.

It can therefore be seen that star formation requires a precise amount of gravity to proceed. This kind of fine-tuning can only be imagined as a function of God. Human beings, or lesser minds than that of God, do not make stars!

12.

Gravity on Planets

The force of gravity is a relatively weak force in nature. The force of gravity is only a tiny, tiny fraction of that of the strong nuclear force that holds the center of atoms together. This relative weakness of gravity is important for living things on planets. For example, even a moderate increase in the strength of gravity on earth would cause any normal-sized person to be crushed. In such a gravity field, even small insects would need very thick legs to move about. In fact, calculations of this kind of situation show that no creature on earth could be much larger than a small insect!

On the other hand, if there were too little or no gravity on earth, it would not be possible for humans to remain fixed to this planet. Humans would most probably drift out into space. Notice, for example, how easy it was for the astronauts on the moon to jump high off its surface. That was possible because the gravity on the moon is only one-sixth that of what it is on the earth!

Gravity on planets, such as the earth, has been fine-tuned. People the size of insects do not exist here! And we are not thrown off the planet, even though the earth turns around at about a thousand miles per hour (at the equator). God has programmed the proper amount of gravity for our development and our safety here on earth.

13.

Polarity of Water

Water (H_2O) is a polar molecule, meaning that its overall charge is unevenly distributed. In other words, it leans toward the negative or minus [-] electrical side of its structure.

H [+1] O [-6] H [+1]

Water has an excessive negative [-] charge near its oxygen atom. Two of oxygen's six [-] electrons in its outer shell are connected to the two [+] charged hydrogen atoms of the H_2O structure. The polarity of the water molecule is caused by the oxygen atom which has four additional free [-] electrons in its outer shell. The ability of other kinds of molecules to dissolve in water is due to these four free electrons [-] of the oxygen atom.

The bottom line is this: The ability of the water molecule to interact in solution with other life-supporting compounds makes water necessary for all living things. Water makes up about 70% of the human body and about 90% of plant bodies!

One final thought. Notice how Jesus walked on the water of the Sea of Galilee (Matthew 14:26) and changed water into wine at the marriage ceremony in Cana (John 2:9). God has power over water, and the fine-tuning of the water molecule for living systems is one more example.

14.

Habitable Zone Around Stars

The center of our star, the Sun, has a temperature of 27 million degrees Fahrenheit and a surface temperature of 10,000 degrees Fahrenheit. Therefore, the light and heat from stars like our sun is great.

Living things on surrounding planets must be in an orbit close enough to their parent star for the necessary light and heat to reach them, but not too close so as to be "burned up." Planets must also not be in an orbit too far away from their parent star, for if they were, water and the various tissues of living things would freeze. Cold is also an important barrier.

Our planet, the Earth, is not too close to the sun so that we will burn or too far away so that we will freeze. However, two planets, Venus and Mercury, orbit the Sun closer than the earth does. These two tortured planets are too hot for life to exist on them. Both Venus and Mercury have a daylight surface temperature of around 800 degrees F.

Orbiting the Sun farther away than Earth are Mars, Jupiter, Saturn, Uranus, and Neptune. These planets are too cold to support complex life. While Mars can get up to 70 degrees F. near its equator during a summer day, its temperature at night will drop to minus 100 degrees F. The temperature in the clouds of Jupiter (Jupiter has no solid surface) will only be about minus 234 degrees F. Saturn's temperature is on average minus 270 degrees F. The overall temperature of Uranus is minus 357 degrees F., while Neptune has an average surface temperature of only minus 346 degrees F.

One can therefore see that, beyond the earth, planetary temperatures are much too cold to support human life. By God's will, however, we upon the earth are in the so called "Goldilocks zone;" a zone that supports life with a just enough light and heat – not too much and not too little!

15.

Magnetic Shield

The "solar wind" is a continuous flow of deadly atomic particles that travel at almost two million miles per hour from the sun toward the earth. These charged particles are primarily the nuclei of hydrogen atoms. The solar wind is released from the upper atmosphere of the sun. Its particles can escape the sun's gravity because of their high velocity. The only time the solar wind can be seen on earth is when it creates the "northern lights," or as it is sometimes called, the "Aurora Borealis."

From time to time, invisible particle surges come from "solar storms," which then affect radio and TV communications on earth as well as navigation systems and satellites. Such high energy magnetic storms are especially noticeable when "solar flares" or "coronal mass ejections" occur on the sun. If the Earth were not shielded by its "magnetosphere," the sun's radiation would seriously damage human DNA and cause various forms of cancer. The Earth fortunately has a large, bubble-like magnetic shield that redirects the Solar Wind around the earth.

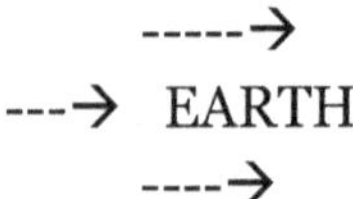

This magnetic envelope is created by the internal structure of the earth. The earth's internal anatomy can be mapped out by studying the way shock waves from earthquakes move through the earth. From this data it is known that the earth has an inner core, composed mostly of iron, that is about two-thirds the size of the moon. This inner core has an interior temperature of about 10,000 degrees F. and is under enormous pressure. This iron core is solid in structure. However, flowing above it and under far less pressure is a 1200-mile-thick layer of liquid iron and nickel. And because the earth spins at 1000 miles per hour (at the equator) this liquid layer moves and generates electric effects. These electric effects in turn create a vast magnetic field around the earth, the magnetosphere, which protects life on this planet from the solar wind.

This solar wind has already driven off most of the atmosphere on the nearby planets of Mercury, on one side of us, and on the surface of Mars, on the other side of us. In other words, these two planets have no oxygen gas above their surfaces as we do here upon Earth. If the same thing had happened on earth, animals on Earth would not have any oxygen to breath! Human life would not have been possible here.

Once again, we "see" the invisible hand of God working in our favor. He sees to it that we have enough light and heat on this planet from the sun, but also designs into the earth a solid iron, liquid interface to generate an invisible electric field to deflect the solar wind. How thoughtful!

16.

UV Shield

Ozone is a molecule consisting of three oxygen atoms hooked together (O_3). Ozone forms within the stratosphere, between 12 and 19 miles above the earth's surface. What happens there is that high energy ultraviolet (UV) radiation from the sun hits an ordinary O_2 (oxygen molecule) and splits it into two parts. Then, these free oxygen atoms collide once again with other O_2 molecules to form the new ozone molecule (O_3).

The ozone shields living things on earth from the *harmful* ultraviolet B radiation from the sun. This short-wave radiation can be a cause of various skin cancers, including malignant melanoma. In addition, UVB has been linked to cataract formation in the human eye and is also the cause of "sun burn."

Ozone is another help that God has given us here on earth.

17.

The Moon and Earth's Climate

The Moon is not just majestic and pleasant to look at. If the Moon were not exactly where it is, the climate upon earth would be chaotic! The Earth-Moon system is really a double-planetary system, which is very rare among possible planetary configurations.

The north-south axis of our earth is tilted about 23 degrees with respect to the sun. This 23-degree tilt has effects. When a specific location on the Earth is tilted toward the sun, it is summer time. However, when a specific location on the Earth is tilted 23 degrees away from the sun, it is winter time.

The presence of the Moon's gravity stabilizes this 23 degree tilt of the Earth and thereby helps to moderate the seasons here on Earth. Even now, with the Moon's present influence, the Earth still "wobbles" at about 1%. Without the Moon's gravity to stabilize the tilt of the Earth, this wobbling would be 10-20%.

In the past, when the earth was tilted only 1% more away from the sun, the "ice ages" occurred. In times like that, winters were longer and colder. One example comes to mind: about 20,000 years ago, New York City was covered by a "sheet" of glacial ice one mile deep!

So, once again, we "see" the invisible hand of God. The rarity of being situated within a double-planetary system, while the earth is tilted to exactly 23 degrees, and the Moon is stabilizing that position!

> Who has known the mind of the Lord?
> Or who has been His counselor?
>
> Romans 11:34

PART TWO:

SOUL IN THE BIBLE

18.

Soul in the Early Bible

The Bible has more than a thousand-year history. From its earliest days, God spoke to the Old Testament prophets, revealing the truth about Himself and about his creatures.

Working within the minds of his prophets, God revealed not only that He was the eternal Necessary Being, but also that humans were created in His own image. In part, this meant that humans would have God's eternal component in their being: "in the image of God He created them" (Genesis 1:26). Since God always lives, so too will His people!

This component of immortality was spoken of in the OT as the "soul." The soul is invisible, like God Himself. But the soul is real, it is there – a "spiritual body" one might call it (1 Corinthians 15:44).

The OT writers were given revelation through the Holy Spirit. They were told that each soul is the holy "breath" of God Himself (Genesis 2:7). When this breath or soul leaves the body, the person's body dies. When an embryo is conceived in the womb, it is implanted with God's living breath. That is why killing any human being, whether inside the womb or out, is very wrong! The belief in the sacredness of life is an important sign of civilized people.

The following OT quotations will illustrate how serious this topic of the human soul is. One sees it very early in Scripture!

> Yahweh God [the Father] formed the man
> from the dust of the ground [his body], and
> breathed into his nostrils the breath of life
> [his soul], and the man became a living soul.
>
> Genesis 2:7

> Rachel brought forth [a boy], and had
> [a very] hard labor in her bringing forth ...
> And it happened as her soul was going
> forth [from her body] that she died ...
>
> Genesis 35:16-18

> Seek Yahweh your God ... with all
> your soul, [even] in your distress ...
>
> Deuteronomy 4:29

> Yahweh your God ... keep his commandments
> and cleave to Him ...with all your soul.
>
> Joshua 22:5

> The soul of Jonathan was
> knitted to the soul of David
>
> 1 Samuel 18:1

Yahweh [God the Father] is my shepherd ...
He restores my soul ...
Psalm 23:1-3

Why are you downcast, O my soul?
And why do you moan within me? Hope in God ...
Psalm 42:11

O God, my God ...
my soul thirsts for you;
my flesh longs for you ...
Psalm 63:1

Bless Yahweh, O my soul,
and within me His holy Name.
Psalm 103:1

Yahweh will keep you from all evil;
He shall keep your soul.
Psalm 121:7

I wait for Yahweh [God the Father];
my soul waits and I hope for His Word.
Psalm 130:5

You [God] have possessed my inward parts;
you wove me in the womb of my mother.
Psalm 139:13

Pleasant words are ... sweetness
to the soul and healing to the bones.
Proverbs 16:24

The spirit [of me] shall
return to God who gave it.
Ecclesiastes 12:7

Yahweh says ... ask ... where
the good way is, and walk in it, and
you shall find rest for your souls.
Jeremiah 6:16

The soul of the father, also
the soul of the son – they are Mine ...
Ezekiel 18:4

19.

Jesus Speaks of the Soul

Jesus treated the soul's existence as a fact. St. Thomas Aquinas said: "The soul is in the body as containing it – not as contained by it."

[Jesus said] Do not fear those
who kill the [physical] body,
but cannot kill the soul ...

Matthew 10:28

He said further that the soul is more real, more certain, than the world itself.

What will it benefit a man if he gains
the whole world, but loses his soul?
Or what will a man give in exchange for his soul?

Matthew 16:26
[Mark 8:36-37]

Jesus spoke of the soul's ongoing existence.

I am the God of Abraham ... of Isaac, and ...
of Jacob [who had lived on this planet
hundreds of years earlier and were still alive].
He is not the God of dead [people] but of living
[ones].

Matthew 22:32-33
[Mark12:26-27]

He [Jesus] said ... "You shall love God
with all your heart, and with all your soul ...
This is the ... first commandment ..."

Matthew 22:37-38
[Mark 12:30]

I will say to my soul: "Soul, you have many goods ...
So eat, drink, and be merry!" But God said to him:
"You fool!
This very night your soul will be required of you ...!"

Luke 12:19-20

I [Jesus] say to you: "Do not be anxious in your soul
about what you might eat, or ... what you might put
on."

Luke 12:22

I [Jesus] am the good shepherd, the good shepherd
offers his soul on behalf of the sheep ...

John 10:11

20.

NT Letters Speak of the Soul

In the NT documents the words spirit and soul are often used interchangeably.

Stephen called out: "Lord Jesus,
receive my spirit [into Heaven] ..."

Acts 7:59

[Paul was] in Antioch ... strengthening
the souls of the disciples ...

Acts 14:21-22

the souls of all men ...

Romans 2:9

The Spirit testifies to our spirit
that we are children of God.

Romans 8:16

on behalf of my soul [that of Paul]
they risked their necks ...

Romans 16:4

I [Paul] was ... absent from the body,
but present in the spirit

1 Corinthians 5:3

that his spirit might be
saved on the day of the Lord.

1 Corinthians 5:5

that she might be holy
both in body and in the spirit ...

1 Corinthians 7:34

The first man Adam was made a living soul.

1 Corinthians 11:45

I [Paul] call on God the testifier
concerning my soul ...

2 Corinthians 1:2

the outside of us [the body] is wasting away,
but the inside of us [the soul] is being renewed
day by day ... [and] is eternal.

2 Corinthians 4:16

I [Paul] would rather be away from
the body and at home with the Lord.

2 Corinthians 5:8

a man [Paul] ...whether in the body or
out of the body [the soul] I know not,
but God knows ... having been seized
[taken up temporarily] into the third heaven
[Heaven].

2 Corinthians 12:2

May the grace of our Lord
Jesus Christ be with your spirit...

Galatians 6:18

in the spirit of your mind put on the new man ...

Ephesians 4:23-24

doing the will of God with the soul ...

Ephesians 6:6

My desire [St. Paul] is to depart [from the body]
and be with Christ [in Heaven].

Philippians 1:23

the Lord Jesus Christ be with your spirit.

Philippians 4:23

hope [is]... an anchor of the soul,
both firm and secure ...

Hebrews 6:19

faith with the possession of [our] soul.

Hebrews 10:39

stay awake on behalf of your souls.
Hebrews 13:17

[Jesus is] the one thing
being able to save your souls.
James 1:2

The body without the spirit is dead.
James 2:26

deliver the soul of him from [eternal] death.
James 5:20

your faith [is] the salvation of your souls.
1 Peter 1:9

Your souls have been purified
by your obedience to the truth.
1 Peter 1:22

fleshly desires war against the soul ...
1 Peter 2:11

return to the shepherd [Jesus] ...
overseer of your souls.
1 Peter 2:25

eight souls were saved through water ...
Baptism now saves you.
1 Peter 3:20-21

the righteous soul, living among them ...
2 Peter 2:8

ensnaring weak souls ...
2 Peter 2:14

the spirit of life from God entered them
[the dead], and they stood up on their feet,
and great fear fell upon the ones watching.
Revelation 11:11

21.

Jesus Reveals Life after Death

This section will briefly examine three events in the life of Jesus: [1] His transfiguration [2] His actions with the Good Thief upon the cross; and [3] His resurrection appearances over a forty-day period.

Sometime before Jesus was arrested, he took Peter, James, and John up a mountain and revealed himself before them in resurrection glory. He became bright in appearance, gleaming white; and, importantly, two persons from Heaven also appeared and spoke with Jesus. These two men were Moses, who had lived about 1450 BC, and Elijah, the prophet, who was alive about 850 BC. These two persons evidence the reality of the invisible existence of the human soul. Their earthly bodies may not have still been intact, but they as persons were fully alive on this hill. They were materialized from Heaven in palpable bodies. In fact, so real were they that Peter offered to make "three tents," one for each of the three persons, as on the Exodus out of Egypt in the time of Moses.

> Jesus took with him Peter, James, and his brother John; and led them up a high hill by themselves. And he [Jesus] was transformed before them, and his face shone as the sun, and his clothes became white as light. Suddenly there appeared to them Moses and Elijah, talking with him [Jesus]. Then Peter said to Jesus, "Lord, it is good for us to be here; if you wish, I will make here three tents, one for you, one for Moses, and one for Elijah." While he was still speaking a bright cloud overshadowed them, and from the cloud a voice said: "This is [Jesus] my Son, the loved one ... hear him." And hearing this the disciples fell on their faces and were very afraid.

Matthew 17:1-8
[Mark 9:2-8 & Luke 9:28-36]

A second event to be studied is the interaction between Jesus and the so called "Good Thief," who was fastened to a cross nearby that of Jesus. The scene of this event is stark. There is darkness over the whole area (Mt 27:48), and an earthquake shakes the land so that graves are opened and dead bodies begin walking out (Mt 27:51-52). The Roman guards were understandably frightened and instantaneously became believers (Mt 27:56). In this monumental moment of history, the converted robber on the cross turns to Jesus and says: "Jesus, remember me when you come into your Kingdom." To this statement Jesus replied: "Truly, I say to you, today you will be with me in Paradise." (Luke 23:43)

The "Paradise" of which Jesus spoke was Heaven.

> To the one conquering, I will give to him to eat ... in the Paradise of God [Heaven].
>
> Revelation 2:7

St. Paul also speaks of this "Paradise" as being in Heaven. There are three "heavens" in the Bible. Heaven #1 is the atmospheric heaven of the sky and clouds. Heaven #2 is the heaven of the stars. Heaven #3 is the Heaven of God and his friends. This third Heaven is the Heaven Paul spoke of when using the word "Paradise."

> I know a man [Paul] ... [who] was seized up into the third Heaven ...he was seized into the Paradise and heard unspeakable words which it is not possible for a man to speak [back here on earth]
>
> 2 Corinthians 12:2-4

Notice especially that Jesus said to the Good Thief that "this day" he would be with Jesus in Heaven. The Good Thief's body would remain on earth, but he would be in Heaven that day! Also, Jesus's body would be buried that day by Joseph of

Arimathea and Nicodemus (John 19:38-41). Nicodemus brought myrrh and aloes to embalm Jesus' body. And a great stone was also used to seal the tomb (Mt 27:60). But Jesus was already in Heaven! Jesus, the second subject of God, would not, of course, be confined to the ground. He would simply rejoin his body on the third day, on Easter Sunday, and rise from the dead (Mt 28:6).

But what about the Good Thief who was not a divine Person? How did he get to Heaven that day, when his body was still on earth and was taken down from the cross (John 19:31-33)? This, then, is another example of the reality of the human soul. The Good Thief's soul, or spiritual body, was in Heaven, while his physical body was still on earth (1 Corinthians 15:44)!

The third topic to be considered is the appearances of Jesus after his resurrection from the dead, over a period of "forty days."

> After his suffering he [Jesus] presented himself alive to them by many convincing proofs, appearing to them during forty days [1 1/3 months] and speaking about the Kingdom of God.
>
> Acts 1:3

The appearances of Jesus, after his resurrection, can teach us something about life beyond death. The first thing one notices is the power Jesus had over his risen body. For example, when Mary Magdalene stood outside the empty tomb on the morning of

Easter Sunday, she did not immediately recognize Jesus as he spoke to her.

> ... she turned around and saw Jesus standing there, but she did not know that it was Jesus. Jesus said to her: Why are you weeping? Who are you looking for?" Supposing him to be the gardener, she said to him: "Sir, if you have carried him away [from the tomb], tell me where you have laid him, and I will take him away." Jesus [then] said to her: "Mary!"
>
> John 20:14-16

Later that day, the risen Jesus walked with two of his disciples from Jerusalem to Emmaus, a distance of about seven miles. During that time Jesus spoke to them about his presence in the Old Testament; yet while doing this, they did not recognize him!

> Their eyes were being held from recognizing him.
>
> Luke 24:16

Mark, however, said that Jesus appeared to the two disciples "in another form."

> After that he [Jesus] appeared in another form to two of them, as they walked into the country [to Emmaus].
>
> Mark 16:12

This theme of "another form" is amplified even further at the end of their walk to Emmaus. Since it was late in the day, the two men "pressed him" (Jesus) to stay overnight with them. Thus, the reader would be shown yet another dimension of his risen life. That evening Jesus would, at table: take bread, bless it, break it, and give it to the two disciples.

These four steps — take, bless, break, and give — were earlier used at the Last Supper when Jesus said: "Do this [action] in remembrance of me." At the Last Supper he also underlined the reality of the Eucharist, by saying: "This [transformed] bread *is* my Body" (Luke 22:19) And so, after the liturgy of the Word (his talk about himself in the OT), the two disciples would be given a view of the "liturgy of the Eucharist" at their table. A complete "Mass" would thereby have been experienced by them.

So, after Jesus had consecrated the bread at supper, he handed it to the two disciples. They now held Jesus (the bread) in their hands, a form of his non-material, non-corporeal manifestation; a form at least as mysterious as his Presence in a risen physical body.

Then, Jesus immediately disappeared before their eyes! He was literally gone from their midst. He was, however, present to them in his new Eucharist form of Being.

> As he [Jesus] sat at table with them, he took bread, blessed it, broke it and gave it to them. And their eyes were opened, and they [finally] knew him; and he [then] *vanished*

> *out of their sight.*
>
> Luke 24:30-31 & 35

Another surprising insight into this "risen state" was Jesus's appearance to his Apostles in Jerusalem, late on that first Easter Sunday. Jesus came to them through a closed door, instantaneously appearing to them within the upper room. At first, they thought they were seeing "a spirit"; but then he showed them his wounded hands and feet and ate a piece of their "grilled fish." He was, therefore, a real person in an empowered body.

> When they were speaking he [the risen Jesus] stood in the middle of them, and said: "Peace be to you." They [the Apostles] were terrified and thought they were seeing a spirit. And he said to them: "Why are you troubled ...see my hands and my feet that I am myself; touch me and see" ... Still un-trusting ... he said to them: "Have you some edible food?" They gave him part of a grilled fish, and ... he ate.
>
> Luke 24:36-43

The "risen state" of ordinary people may not exactly resemble that of the risen Jesus, but it is an example to consider. As a "risen soul," no longer in a physical body, we might appear in a form like that of the "spirit" the Apostles thought they were seeing. A "spiritual body," to which St. Paul would later make reference to in 1 Corinthians 15:44, has no "mass" or physical substance. But neither do the "photon" particles that make up the "light" we see around us! And neither do the "gluon particles" within the nucleus of all atoms. They too are "massless," or "no substance," particles! So, the mystery of the "spirit" experience, and that of "massless particles," goes forward; something for a prayerful soul to meditate upon.

Jesus's appearances continued over a 40-day period, before he "ascended," or became more permanently separated from the "sight experience" of ordinary people. St. Paul, twenty years after Jesus' resurrection, reviewed a few of his earlier appearances.

> he [the risen Jesus] appeared to Cephas [Aramaic for Peter], then to the Twelve [Apostles]. Then he appeared to more than 500 persons at one time, most of whom are still alive, though some have fallen asleep [died]. Then he appeared to James and other disciples; and lastly [some time later] ... he was also seen by me.

1 Corinthians 15:5-8

Jesus' "Ascension" from this planet resembled, in some sense, his disappearance before the two men at Emmaus. He was simply "lifted up," and then disappeared within a ground-level "cloud."

Acts 1:9 & Mark 16:19

22.

Jesus: Recent Appearances

After Jesus's Ascension into Heaven, his regular bodily appearances on earth stopped. He was, of course, invisibly Present in the Holy Eucharist of Mass but was not ordinarily present to our earth-bound eyes. We could no longer see him in the ordinary sense.

Jesus has, nonetheless, continued to appear to select individuals throughout history. As a divine Person, unlike ordinary humans, Jesus still remains a witness to life beyond death and to the reality of the invisible human soul.

In this section, four such appearances of Jesus will be briefly discussed. First up, Jesus's appearance to St. Francis of Assisi.

[1] Someone has said that a leader is someone whom others will follow. St. Francis of Assisi (1181-

1226) was certainly a leader during his short life. As a young man, he was handsome, intelligent, witty, liked fine clothes, spent his father's money freely, displayed an interest in weapons and was popular with the young nobility around Assisi. He planned a military career for himself and set out with others to attack the city state of Perugia when he was about age 20. He was, however, captured and put under arrest and kept in Perugia for about a year.

In 1205, when Francis was 24, he stopped one day at a rundown chapel known as San Damiano (St. Damian) in Assisi. While praying there, the crucifix before him became alive, and Jesus spoke to him, saying to him three times: "Go, Francis and repair my house which, as you can see, is falling into ruins!"

Following this live, speaking revelation, Francis went door to door in Assisi begging for stones to restore St. Damian chapel. Francis laid the stones himself to restore the chapel.

He also restored another rundown chapel on the plain below Assisi, called, "St. Mary of the Angels," also known as the "Porziuncola." Here, in 1216, Jesus appeared to Francis a second time, offering to help in this work.

Francis, like Jesus, chose poverty as a way of life. Soon others joined him, and, at the age of 28 in 1209, he drew up a rule for a new community. With 11 followers, he went to Rome and managed to obtain Pope Innocent III's approval. St. Mary of the Angels church became the base for their new religious community, and by 1220, when the order met there as a whole, there were already 5000 Franciscan friars and 500 applicants! The community had also spread to three other countries.

In the year 1222, Francis received the five wounds of Jesus on his hands, feet and side, also known as the "stigmata." Francis lived with these bloody wounds until his death in 1226 at age 44. He was declared to be a "saint" in 1228 by Pope Gregory IX.

[2] A second person to whom Jesus appeared was St. Catherine of Siena (1347-1380). When Catherine was only six years old, she went along with her older brother on a visit to an older married sister. On the way home she suddenly stopped still on the road and gazed up into the sky. There she saw a vision of Christ seated in glory with the Apostles Peter, Paul, and John around Him.

In 1366, when Catherine was 19, she had another vision of Christ. It was during carnival time in Siena, Italy. Catherine was praying in her little 9-x-3-foot room. Suddenly Christ appeared to her, accompanied by the blessed Virgin Mary. Mary took Catherine's hand and held it up to Christ, who placed an invisible ring upon her finger. To Catherine, this ring was always visible, though others could not see it. After this event, Catherine used her time to serve the poor, even those suffering from the bubonic plague. A group of spiritual friends gradually gathered around her. They included many priests and religious women.

In February, 1375, at age 28, Catherine visited Pisa, Italy, where she had yet another life-changing experience. She had just received communion in the little church of St. Christina and was praying before a crucifix, when suddenly three blood-red light rays approached her and, in her humility, she asked that they not scar her. They changed to white light rays, which pierced her hands, feet, and side causing considerable pain. These wounds, the stigmata of Christ, remained with her until the end of her life. They were not visible for others to see, but when she died, they were clearly present on her body.

Catherine also wrote almost 400 spiritual letters during her short lifetime and an important book – later given the title: *The Dialogue of St. Catherine.* In 1376, she was called to Avignon, France, to meet with Pope Gregory XI concerning hostilities between certain Italian cities. While there, she persuaded the pope to move the Church offices back to Rome, after having been in Avignon for 73 years.

Catherine died at age 33 in 1380. Here are three quotations of St. Catherine:

Speak the truth in a million voices. It is silence that kills.

Be who God meant you to be and you will set the world on fire.

Sail with the wind that Heaven sends us.

[3] A third example of communication with Jesus after his Ascension is that of St Julian of Norwich in England. St. Julian resided in England

between 1342 and 1414. When she was thirty years old, she acquired a serious illness that brought her close to death. A priest was called in on May 8, 1373, to give her the last rites of the Catholic Church. As part of the prayer service, the priest held up a crucifix above the foot of Julian's bed. As Julian looked at the crucifix the figure of Jesus on the cross began to bleed. This sighting of the suffering Jesus continued for the next five days. The visions of Jesus ended on May 13, 1373.

Julian wrote about these visions immediately after they happened. Her work of 25 chapters is titled "Revelations of Divine Love." What follows are three quotations from Julian's "Revelations."

As the body is clad in cloth ... so are we,
soul and body, clad in the Goodness of God.

God is nearer to us than our soul. For he is
the ground in which it stands ...so if we want
to know our own soul, and enjoy its fellowship,
it is necessary to seek it in our Lord God.

The greatest honor we can give Almighty God is
to live gladly, because of the knowledge of his love.

[4] St. Padre Pio (Francesco Forgione) was born in 1887 in a farming village in southern Italy called Pietrelcina. In 1903, when he was 16 years old, he entered the Franciscan community to study for the priesthood. It was during the years 1903-1904 that Padre Pio experienced his first visions of Jesus and

Mary, but they were not reported to the public. Later, after seven years of study, he was ordained a Catholic priest.

In 1911, the five wounds of Christ, the stigmata, appeared on his body for the first time. He was 24 and very disturbed by this. He prayed that the Lord would withdraw the stigmata. He did not ask for the pain to be removed but only the visible wounds. The visible wounds did, in fact, disappear for a time, and Padre Pio was assigned to the mountain community of San Giovanni Rotondo in 1916. Padre Pio remained there for the rest of his life.

In July, 1918, Pope Benedict XV asked for all Christians to pray that the First World War (1914-1918) would come to an end. In response to his call, on July 27, 1918, Padre Pio offered himself as a sacrifice to God to end the war. Then on August 5, 1918, Jesus again appeared to Padre Pio, and part of the pain of the stigmata returned. Finally, on September 20, 1918, the wounds of Jesus, the stigmata, again returned in full. These five wounds on his hands, feet, and side remained with the priest for the next fifty years, until his death in 1968.The blood flowing from these wounds had the scent of flowers, and the wounds never became infected, a fact that amazed more than one medical doctor.

Padre Pio was also known for various miracles that occurred around him and had the reputation of being able to read souls in the confessional. On one occasion, his only food for a period of twenty days was the Eucharist. Fr. Agustino, of Padre Pio's community, was a witness to this occurrence.

23.

Mary: Recent Appearances

The appearances of the divine Jesus to "saintly people" on earth might be expected from time to time. Mary, however, is another story! Mary was like us! She was, and is, a totally ordinary human person. She was not, and is not, divine in any way. She was, however, the carrier of the only incarnate Son of God for this planet. And, as such, Mary was preserved from original sin in a way that we are not.

I [King David] was brought forth in iniquity [sin] ,
and in sin did my mother conceive me.
Psalm 51:5

One probable reason Mary remained free from sin is expressed in Job 14:4:

none can bring a clean thing [Jesus]
out of an unclean thing [Mary].

The Bible witnesses to the fact that Mary is in Heaven. The Book of Revelation lays out the story. There is:

> seen in Heaven a woman clothed with the sun, with the moon under her feet, and on her head was a crown of twelve stars ... she gave birth to a son ... who is to rule all nations [Jesus] ... the dragon [Satan] went off to make war on the rest of her children, those who keep the commandments of God and hold the testimony of Jesus.
>
> Revelation 12: 1;5;17

Mary is therefore with the Lord; and like the angels, has become a carrier of God's messages back here upon earth. This section will examine two examples of Mary's appearances on this planet, bringing messages from God to his people still on earth.

[1] Mary in Mexico

The first appearance of Mary we will examine is her appearance to Juan Diego outside of Mexico City in 1531. The site of Mary's appearance to Juan is presently inside the city proper, where the Basilica of Our Lady of Guadalupe is now located.

In 1369, the town of Tenochtitlan (today's Mexico City) was founded. The Aztec people who resided there were led by leaders dedicated to war. Their warriors would conquer nearby peoples, forcing them to provide tribute and additional men to conduct war, plus bodies to sacrifice to their sun god.

In 1487, a large temple was built in Tenochtitlan to provide sacrifices for this sun god. Thousands of human lives were forcefully taken inside this temple to satisfy the sun god, who was believed to be in constant battle with the moon and surrounding stars. It was thought that the sun god needed human blood to maintain his strength. The Aztec priests further taught that if the sun god died all life on earth would end. Under such beliefs the Aztec people lived in fear.

Mary, carrying out the intentions of Jesus, penetrated this Aztec culture to free its people. Thus, she would become the principal evangelist of the Mexican people.

In 1474, Juan Diego was born. When little, Juan's father died, and Juan was taken in and raised by his uncle, Juan Bernardino. Later, in 1519, when Juan Diego was 45 years old, Cortez sailed from Spain with 500 soldiers and entered Mexico on horseback.

Then, in 1524, the first Catholic priests, twelve in number, came to what is today Mexico City. They were very kind to the Aztec people, in sharp contrast to the fear-generating Aztec priests. As a result of their kindness and instruction, Juan Diego and his wife, Maria Lucia, were among the earliest of the Aztec people to be baptized. Juan's uncle, Juan Bernardino, also became a Christian.

In 1529, Juan Diego's wife died, and Juan left the

town of Tenochtitlan (now Mexico City) and moved back in with his uncle about seven miles away. Two years later, on December 9, 1531, Juan Diego set out early on a Saturday morning to attend Mass at the Franciscan church near Mexico City. As he walked past Tepeyac Hill, he heard a gentle song that seemed to be coming from some birds. However, when he looked up to the top of the hill, he saw a white, radiant cloud. Then the song stopped, and he heard the voice of a woman calling him: "Juanito!" So, he climbed up the hill and saw a young, attractive woman who spoke to him in his native "Nahuatl" language. Her clothes shined like the sun. She asked Juan where he was going, and after Juan answered, she said:

> I want you to know for certain, my son, that I am ... the virgin Mary, mother of the true God ... I greatly desire that a church be built here in my honor ... I will comfort your people's affliction and their suffering ... Go now to the bishop in Mexico City ... make known to him the great desire I have to see a church here.

Juan Diego went directly to see bishop Zumarraga, the first bishop of Mexico, a pious and very careful man. Bishop Zumarraga listened carefully to Juan but thought that Juan had experienced an illusion and therefore did not believe his story. Then, when Juan was returning home, he again encountered Mary at Tepeyac Hill and told her about his experience with the bishop, and then added: "I beg you to entrust your message to someone more known and respected [than me], so that he will believe it." Mary rejected Juan's plea and asked him to return to the bishop the next day, Sunday: "Tell him that it is I myself, the Blessed Virgin Mary, mother of God, who am sending you."

On Sunday morning, after Mass, Juan Diego again went to the bishop's house. The bishop again listened attentively to Juan and asked questions but said he would need a more tangible sign to prove that it was really Mary speaking. Then, when Juan left, the bishop discreetly sent two servants to follow him. However, when Juan reached Tepeyac Hill, he disappeared. The servants could not find him! At that very time, Juan was speaking to Mary who had been waiting for him on the hill. She told Juan to: "Come back tomorrow morning, to receive the sign he [the bishop] is asking for."

When Juan arrived at his uncle's house, he found the man very sick. The next day Juan did not go to the hill but stayed with his uncle to care for him. Then on Tuesday, December 12th, his uncle asked Juan to go seek a priest for the last rites of the Church.

On his way to Mexico City, Juan intentionally tried to detour around Tepeyac Hill so as not to encounter Mary. She, however, came down to meet him. He was embarrassed, and, while explaining the situation, said that after finding a priest and giving his uncle the last rites, he would return. Mary then replied:

> My dear little one ... do not be distressed about your uncle's illness, because he will not die from it. I promise you that he will get well ... Now go to the top of the hill, pick the flowers you will see there, and bring them to me.

When Juan reached the top of Tepeyac Hill, he was stunned to find a great number of flowers in bloom: Castilian roses in the middle of winter! This was important because Castile was a province in Spain, the bishop's home country. Juan gathered the roses, holding them in his *tilma* (cloak) and then came down the hill to Mary. Mary arranged the roses with her hands, and then said: "My dear son, these flowers are the sign that you are to give the bishop ... This will get him to build the church that I have asked of him."

Juan Diego went directly to the bishop, but the bishop's servants kept him waiting for some time. Nonetheless, when Juan finally saw the bishop the flowers still had dew on them. When Juan dropped the flowers to the floor, tears immediately entered the bishop's eyes and he knelt down because an image or portrait of Mary was etched onto the *tilma* that Juan was wearing. It was the picture of "Our Lady of Guadalupe," as can still be seen in all its freshness within the Basilica in Mexico City.

After placing the roses and *tilma* with its portrait in a special room, Juan and the bishop went to Tepeyac Hill. The bishop reviewed the site and then sent Juan on his way to care for his uncle. When

Juan arrived, he found his uncle Juan Bernardino completely well again. His uncle said that Mary had appeared also to him, saying that she wanted the church built, and that her portrait was to be called, "Saint Mary of Guadalupe." (This term, "Guadalupe," came from Spain; a very old sanctuary there is dedicated to Our Lady of Guadalupe.)

A year later, on December 25th, Bishop Zumarraga led a procession to Tepeyac Hill where a new sanctuary to Mary was dedicated. Juan Diego moved in next to this site and farmed a nearby plot of land until he died on December 9, 1548, 17 years to the day after the first appearance of Mary.

Mary kept her promise, given to Juan in 1531. By 1540, nine million people in Mexico had been baptized, and human sacrifice was ended. About 3000 souls a day were becoming followers of Jesus.

The portrait of Mary on the *tilma* showed her to be more powerful than the Aztec sun god. In the image, Mary appears to be standing before the sun. She is also shown to be greater than the moon since her feet are placed upon it. And she is no longer situated only in this world, as she is shown above it with an angel. And, lastly, her hands are folded in prayer, acknowledging that God is greater even than she. The image of Our Lady of Guadalupe also matched the image provided in the Book of Revelation 12:1.

The *tilma* upon which the holy image is situated is made of cactus fiber – a plant material. The image is about 4 ½ feet (1. 43 meters) tall. The fact that this image is perfectly preserved after almost 500 years is unexplainable. The garment appears new!

In 1936 two fibers were examined from the *tilma*: one red and the other yellow. No known coloring agent could be found for either color. There is also no trace of a sketch under the image, and no background paint (a "primer") to prevent the portrait colors from being absorbed back into the fabric. There are also no brush strokes on the image that can be detected. In the portrait, Mary's eyes reflect the image of a man, probably bishop Zumarrraga. Also, on Mary's neck is a brooch, the center of which contains a little cross, recalling the death of Jesus.

[2] Mary in France

In 1858, Mary, the mother of Jesus, possessing a human soul like ours, appeared to St. Bernadette at Lourdes in France. Mary appeared several times over the period from February 11 to July 16, 1858. She appeared as a real person, visibly seen. As Bernadette said to Fr. Peyramale: "I see her as clearly as I see you talk to me." During Mary's third appearance, Bernadette reported the following: "The Lady came down from the niche and stood beside me ... and she laughed."

Mary's presence was not as an invisible soul, but rather as a vibrant, responsive person. She had a "spiritual body" (1 Corinthians 15:44).

Each time she appeared, there seemed to be considerable white light around her. Of Mary's first appearance on February 11th, Bernadette had the following to say: "The Lady was dressed in white ... a blue sash at her waist ... [and] on that cold winter's day her feet were bare ..."

White light, or its composite colors, consists of massless atomic particles known as "photons." Photon particles have no mass or measurable substance but manifest themselves as "light." Mary's body, a spiritual body, may have been expressed by way of such photon particles.

The appearances of Mary to St. Bernadette has been variously recorded. Bernadette herself gave an account. Some of the details from her account will be here presented.

On Thursday, February 11, 1858, Bernadette Soubirous (aged 14) set out with her sister Toinette (age 11) and a friend, Jeann Abadie (age 12), to gather firewood for their families. There was a small mill stream near the River Gave in Lourdes. The two younger girls crossed the stream, while Bernadette sat down to take off her shoes before crossing. Suddenly, she heard a "noise like a storm." She looked in the direction of the noise across the tiny stream and saw a cave in the rock face that later was called "Massabielle." As she looked, "a cloud came out of the cave and flooded the niche with light. Then a lady, young and beautiful, stood on the edge of the niche. She smiled at me and beckoned me to come closer." Bernadette went down on her knees, took out her Rosary, and began to pray. Mary had a Rosary on her arm and followed Bernadette by fingering the beads as Bernadette said each of the prayers. When the Rosary prayers were completed,

Mary smiled at Bernadette and bowed, and then disappeared. This was the first of the appearances of Mary to Bernadette.

The following Sunday, during Mary's second appearance, the grown son of a Lourdes resident picked Bernadette up and forcibly carried her away from the grotto to her house. In this situation, Bernadette said, "the lady kept in front of me and slightly above me!" as they moved toward her house.

In the third appearance, Mary "stood beside" Bernadette and even "laughed." When Bernadette asked the lady's name, Mary instead requested that Bernadette "tell the priests that a chapel must be built here." Then Mary added the following: "I do not promise to make you happy in this life, but [rather] in the next."

This statement is interesting on several points. First, Bernadette is told that there is life beyond death for her. Implied in this is the existence of her soul and a life for it in Heaven. Second, Mary, not Jesus, is doing the promising. This shows how much influence Mary has with Jesus in Heaven. Third, Bernadette died in 1879. Her soul left her body at the moment of her death. Thirty years later, in 1909, when she was being considered for sainthood, her buried physical body was reexamined. It was found to be incorrupt (not affected with decay). Her body, in other words, looked as if Bernadette had just died – after thirty years!

By the sixth appearance of Mary "hundreds of people were kneeling in the grotto." During this appearance, "she [Mary] looked out over my head and sorrow overshadowed her. I asked her why? She answered: 'Pray for sinners.'"

Before Mary's seventh appearance, Bernadette said that "my soul called me to the grotto ..." This was not a normal hearing, using her physical ears, but rather a hearing of her soul. Then, again on April 7, before Bernadette went to the grotto to see Mary, she told of a similar experience, saying: "the call came clear [to me] in my soul."

In the appearance of Mary on February 25, 1858, Bernadette was told to: "Go, drink at the spring and wash in it." When Bernadette investigated the "spring," she found it to be a small, muddy hole. She could not drink from it, and when she tried to wash her face it became covered with mud. "The people laughed in the morning when all they saw was the mud." However, when she returned in the afternoon with a friend, the mud hole had become a fully functioning spring. "Water was bubbling from the hollow." Of this water, one Louis Bouriette took note. He asked his daughter to bring him some of the water. Louis's right eye had been earlier injured and was currently of little use. However, after washing his eye with the spring water, he declared the next day: "I am cured!" Of this and other cures Bernadette said: "The people who scoffed when I [first] washed in the spring eventually treasured its water as a grace from Heaven."

On March first, another such miracle occurred in Mary's new spring. Catherine Latapie plunged her useless hand into its water and her two paralyzed fingers on her right hand began to work again.

Then, on March second, Mary again requested that Bernadette ask the priests "to build a chapel," and said further to "ask the people to come to the grotto in procession."

On March fourth, about "8000 people gathered around the grotto." On this day Mary gave Bernadette a brief look at Heaven. "The Lady came and lifted me into a world where the language is prayer and the environment is Heaven."

Then, three weeks went by when Mary did not appear to Bernadette. During this time the two-year-old son of Croisine Bouhohort was dying, "His little coffin was already being made." Croisine, however, took the dying child to Massabielle and ... immersed him in the cold spring water. The next day, little Louis was walking around full of life! Doctor Vergez ... and Doctor Dozous officially stated that the child's cure "could not be explained by medical science."

On March 25th, Mary told Bernadette who she was. First, however, they prayed the Rosary together. "Then, when the Lady came very close to me ... I said: 'Mademoiselle, would you be so kind as to tell me who you are, if you please?' ... The Lady leaned tenderly towards me and said, 'I am the Immaculate Conception.'" After these words, Mary disappeared. Bernadette immediately ran to tell Father Peyramale of Mary's identity.

"The good priest stood there stunned ... 'Do you know what this [Immaculate Conception] means?' I said no! 'Then, how can you say the words if you do not understand them?'"

I repeated them [to myself] all along the

way.

I then decided to ask Mademoiselle Estrada the meaning ... she explained how Pope Pius IX had applied these words to Our Blessed Lady [Mary] about four years ago ...[so I concluded] she [Mary] was the Mother of God, and had been stepping out of Heaven to share her soul with me!

Sunday, June third was the day of my First Holy Communion ... My soul had been prepared for Jesus by his Mother ... In Holy Communion I am heart to heart with Jesus.

On July 16th [1858], I was kneeling in the quiet of the parish church ... Suddenly my soul stirred with an impulse now familiar: the Mother of God was calling me ... I hurried to the meadow ... I knelt down and began my Rosary ... This would be the last time I would see her on this earth. I knew, because my Lady had prepared my soul for Jesus [in Holy Communion] and now she would give me to him.

Since that time in 1858, tens of millions of people have visited Lourdes, especially with their sick loved ones. Many miracles have followed upon the examples detailed here. And, importantly, the world has seen the vibrant presence of an ordinary person come out of Heaven to interact with people still here on earth.

Part Three:

The Soul and Knowledge

24.

Soul as the Form of the Body

The human soul is "spirit." The soul is not physical or material in any way. It does not consist of matter, nor does it have parts that can be examined with an electron microscope or analyzed in a laboratory.

The soul, as a spirit, occupies the whole of the human body. If someone were to lose a finger, for example, it could not be said of them that they had lost that much of their soul.

One of the organs that the soul occupies is the brain. The "mind" is an expression of the soul's presence in the brain. The mind is not an organ of the body, or a measurable part of bodily functions. The mind is not the brain, but often interfaces with and even engages the brain. People will sometimes say, "Use your brain!" Or they will say, "Use your head!" The important question about these sayings is this: Who is the "your" in "use your brain"? The

user of the brain is the inner "I," or soul of the person.

When a person dies, the brain dies, but the soul continues to live. The soul has no parts that can disintegrate or "return to dust." The soul simply leaves the body and goes before the Lord in Heaven. In the case of a "near death experience," where the patient is "brain dead," the soul temporarily leaves the body. Later, when the body has been resuscitated, the soul returns to the body, and the person is back! Such cases will be examined in a later chapter.

> The unity of soul and body is so profound that one has to consider the soul to be the "form" of the body ... it is because of its spiritual soul, that the body made of matter, becomes a living, human body ...
>
> "Catechism of the Catholic Church" 365

> [T]hen the Lord God formed man of dust from the ground [the body], and breathed into his nostrils the breath of life [the soul], and man became a living being.
>
> Genesis 2:7

25.

Aquinas on Knowledge

St. Thomas Aquinas (1224-1274) taught that cognition is an activity that the subject performs, not something that happens to him as he watches. In cognition, or knowing, the subject in some way becomes the thing known in such a way that he does not cease to be who he was before. In "knowing," the person intentionally becomes the object known and thus acquires a new perfection.

For Aquinas, knowledge consists of acquiring the "forms" of the things we know, and thereby in some mysterious way becoming one with them. Knowledge involves a kind of union. To become the thing which is known a person must possess it. This possession of the object is not a physical possession of it; rather, it is a mental possession of the "form" of the object. The "form" of the object is what makes the object what it is.

> Intelligent beings are distinguished from non-intelligent beings in that the latter

> possess only their own form; whereas the intelligent being is naturally made to have the 'forms' of other things. The idea [or form] of the thing known is acquired by the knower.
>
> Aquinas: *Summa Theologica* I,14,1

The knower thus "becomes" the accumulator of knowledge, as he possesses the "forms" of various objects. The objects themselves, nonetheless, retain their own identity. If a union involved an exchange in substance, as when a person eats, the "food" thus taken in would lose its identity and become part of the living organism that ate it. The knower, however, without losing his own physical being, becomes identical with the object known because he possesses the "form" of the object known. The knower and the object known are physically distinct, while they are cognitively identical. The knowing being is not limited to its own being but is capable of growing cognitively, while the non-knowing being is limited to its own being. The knower has, in addition to its own "form," the "forms" of various other things known.

> Non-intelligent beings possess their own "form," whereas the intelligent being is naturally adapted to have also the 'forms' of other beings; for the "idea" [or form] of a thing or things known is in the knower. Thus the nature of the non-intelligent beings is more contracted and limited; whereas the

> nature of intelligent beings has a greater range and extension ... it is clear that the immateriality of the things [to be known], is the reason why they can be known ...
>
> Aquinas: *Summa Theologica* I,14.1

"Forms" are made finite in matter. Forms originally existed (and exist) in the mind of God, and when they are united to the matter they "in-form" a particular thing. Things such as "plants" are given a form by God but cannot receive other forms and therefore have no power of knowledge. In the physical world, only one "form" can exist in "this" or "that" material thing. However, in the world of knowledge, the cognitive power (a power of the soul) can receive many "forms" abstracted from individual material objects, and thereby increases the "being" of the knower.

26.

Consciousness

There exists a specific knowledge of one's own soul within us. We recognize our own self as the interior person acquiring knowledge. Consciousness is the awareness of one's own self as the interior "I," or the person active within the body. The soul apprehends the object(s) of knowledge ("forms") and at the same time recognizes itself as the interior person doing this work.

As a conscious human being, each person has control over what is put into their own mind, and how that information is handled, whether responsibly or not. Individual freedoms would not be possible if one were pre-programmed and controlled solely by DNA or brain chemistry. If one were controlled by a machine-like brain, then choice would be minimal or not possible at all. We would operate mechanically, like robots. However, a person does have choice and can search for the good with one's soul or mind. Brain chemistry is certainly an important assistant in this overall process, but essential decisions are made by the spiritual person with a soul.

The brain itself may be viewed as the "computer" of the soul. The soul, on the other hand, is the software directing the brain. The two are not opposed to one another; they simply have different and yet complimentary roles to play. Some forms of thinking, such as pouring a glass of water, may be primarily bodily doings. In fact, all thought may involve some kind of feeling and/or sensing operations of the brain.

When a computer finally breaks down and dies, its software will nonetheless continue to exist. When the body breaks down and dies, the invisible soul (the interior software) will also live on to see a new day – hopefully in Heaven.

27.

Development of Thought

The neurophysiologist and noble prize winner, John Eccles (d.1997) created a diagram to highlight the differences between the brain and the mind. In a simplified version of his structure, which is shown below, the word "soul" is associated with the presence of self-consciousness.

In self-consciousness, people are not only conscious of an object that is before them, such as a "finger," but they are conscious of their own consciousness of that object. In other words, a person is aware of his or her own awareness as well as being aware of the object before them. The soul is fundamentally that interior "I," or the self-awareness of each person, as he or she goes about in the world.

Early Animals	Advanced Animals	People
1. Physical World	Psychic World	Cultural World
2. Sensory Brain States	Perceptual Brain States	Conceptual Brain States
3. Sub-consciousness	Dawn of Consciousness	Self-conscious Soul
4. Neuron Determinism	Trial and Error	Education: books, science, history, art, etc.

The kind of ideas (that is, "sense perceptions") that dogs, cats, and chimps have are equated with individual images. Humans take such images and turn them into abstract ideas, or what are called "concepts." Professor John Eccles believed that this high-end work is done by the soul, which is at least partially responsible for conceptual knowledge. He did not think that conceptual ideas are reducible solely to an efficient synaptic system within the brain.

Most words that people use are not perceptual in

nature. Most human language involves words involving concepts rather than those relating to sensory perceptions. The soul organizes these various ideas within the brain into categories, such as: who, what, when, where, why, and how. Eccles believed that such an ordering procedure *cannot* be explained solely by the efficiency of neuron communications within the brain.

28.

Invisibility of Thought

Have you ever seen a thought? Have you ever seen a picture of a thought? Thoughts are invisible, just like radio and television waves. Thoughts do not have weight, shape, size, or form. Therefore, the question might be reasonably asked: Do thoughts occupy space?

Scientists have never seen a thought with an electron microscope or captured one in a test tube of some laboratory.

Recently, a twelve-year-old boy, with very superior math ability, was placed in an MRI machine and given several number problems to solve. The researchers found that when he was working on these problems, certain areas of the parietal lobes of the boy's brain lit up with considerable activity. The total area involved was about two-times larger than that seen in the brains of average, or less able, math students.

This MRI experiment was followed by the claim of a few researchers that they were "seeing the thought process" used by the boy to solve math problems. However, this claim was not true. If, for example, one were to enter the boy's brain and examine his parietal lobes, all one would "see" are a few thousand nerve cells. This may essentially represent the physical or material interface between the brain and the immaterial soul or mind of the subject.

The nerve cells that one can see in the parietal lobes would not provide a clue as to the thoughts necessary to solve the math problems. In truth, only the mind of the boy himself had access to the necessary invisible thoughts. No one else could either see, or know, of these "in-form-ative" thoughts. The boy was, as is every person, an entirely private first-person witness to his own thoughts and consciousness.

The brain is a material instrument. However, some of the topics or "ideas" of thought have little or nothing to do with materiality and are therefore a sign of the soul's presence within the body. Topics that appear to be essentially non-material in nature are the following: belief, hope, prudence, temperance, fortitude, intentions, and even aspects of justice. These non-material topics immediately come to mind, but there are many others. Topics, like the above, are essentially separated from matter and, therefore, have no weight, shape, size, or form. They are essentially matters of the soul.

29.

Imagination

Inventions seem to appear out of nothing! Some inventions are totally new, meaning that they have never existed previously in the world. Others are based, in part, upon existing knowledge yet are a decided leap forward in complexity and usefulness.

In both cases, however, there is something totally new before us. Where did that "something" come from? Some would answer speedily and speak of "imagination."

But what is imagination and how does a person acquire it? One can imagine, for example, that the soul is involved since some inventions have never previously existed in this world.

Here are a few quotations that explore the meaning of the word "imagination."

"Imagination is the eye of the soul."

Joseph Joubert

"Imagination ... is God's self in the soul."

Henry Ward Beecher

"Imagination is the deepest voice of the soul..."

Pat B. Allen

"Imagination is more important than knowledge. Knowledge is limited to all we [already] know and understand ..."

Albert Einstein

"The imagination ... soars higher than nature does."

Henry David Thoreau

"Imagination is an almost divine faculty which ... perceives ... the secret and intimate connections between things ..."

Charles Baudelaire

30.

Dreams

Sleep is a sign of death. Dreams are a sign of life beyond death. "Rise, the one sleeping, and stand up from the dead ..."

Ephesians 5:14

A person sleeps about one-third of his or her life. If one lives to age 90, he or she will have slept thirty years! During this time a person is essentially unconscious to the world. However, in dreams he or she *is* conscious. The unconsciousness of sleep is a function of the body. The consciousness within our dreams is possibly a function of our soul. The soul is ever alive! While sleep is a kind of death, dreams are a kind of new life. A dream is a succession of images that convey some kind of story. In sleep, the soul may leave the body to discover unseen matters; it may go out for a "night on the town" or into some kind of adventure that God has planned for it. When you are asleep, the soul appears to be the real you. The body may provide a physical container for the soul, but the soul appears to take control during sleep.

Dreams occur primarily during a certain period of sleep. During this period there are rapid eye movements (REM). In an eight-hour sleep experience, most dreams will occur during about two hours of REM sleep. The length of such dreams can vary from between a few seconds to about 30 minutes. The average person has roughly three to five such dreams each night.

> We are such stuff as dreams are made on,
> and our little life is rounded with a sleep.
> Shakespeare: *The Tempes*t, act 4, scene 1

Dr. Eben Alexander, a neurosurgeon who taught for 15 years at Harvard Medical School, relates the following account in his book, *Proof of Heaven* (2012). Dr. Alexander had a young female patient with a dangerous brain tumor, which he surgically removed. The girl's mother later told Dr. Alexander the following story about the girl's dad who had died sometime before her operation.

Early in their (the parents') relationship, while they were dating, the girl's mother had given her future husband a yellow shirt and a fedora to wear with it. A fedora is a soft, usually felt hat, with a curled brim and a crown that is creased down the middle, front to back. The hat gave its wearer a "tough guy look." The mother said that the shirt and hat episode ended when their luggage was lost after a flight during their honeymoon. The yellow shirt and the hat were gone and were never found or again discussed within the family.

The daughter, however, had a dream before she underwent her risky brain operation. The deceased father appeared to his daughter in her dream. He told her that there was nothing to fear in death if she should die. The unusual fact about the daughter's dream was the following: The father was wearing a yellow shirt and a fedora hat when he appeared to his daughter in the dream. The mother was deeply moved by this incident because the combination of a yellow shirt and fedora hat was completely unknown to anyone but the mother and her deceased husband! It had been a lover's secret.

31.

Free Will

Wilder Penfield (1891-1976), a Canadian neurosurgeon, treated serious cases of epilepsy. While operating on a patient's brain, he used electrical stimulation to locate the originating site of the epileptic seizures. The patient was awake because the brain has no nerves that sense pain. Local anesthesia was used only to prevent pain from the opened scalp tissue.

While mapping the patient's brain, Dr. Penfield located a number of sites within the cerebral cortex that controlled various bodily senses and movements. He also discovered that these movements were separate from the patient's "will" to move. In other words, Dr. Penfield found that he could cause the brain to trigger movement of a finger without the patient's "will" being involved. Thus, the patient's "will" was observed to be separate from the work of the brain.

Dr. Penfield stimulated different regions of the

cerebral cortex in his patients, who remained fully conscious. The discoveries went something like this. He would apply a weak electrical current to the "hand" region on the left side of the cerebral cortex. When he did this the patient automatically moved his right hand (on the opposite side). Then, when Dr. Penfield asked his patient why he had moved his hand, the patient said that he himself had not caused the motion, but rather that the motion had been caused by Dr. Penfield's electrode or electrical contact.

When Penfield stimulated the brain area that controls or activates the patient's voice box (larynx), the patient uttered a syllable. When asked about this speech sound, the patient said that he had not uttered it. The patient said that Dr. Penfield's electrical stimulation was responsible for the utterance.

From these and many other like investigations, Dr. Penfield concluded that the patient's "will to move" was separate from the movement of any specific body part; that the conscious "will to move" comes from something other than the brain that is causing the movement. In this sense, the brain might be compared to a computer while the operator of the computer might be compared to the soul or the "will" of its owner.

Another neuroscientist of importance in this discussion of the "human will" is Hans Kornhuber (1928-2009) of Germany. Dr. Kornhuber studied the electrical activity of the cortex of the human brain before, during, and after an intentionally "willed" movement of some kind. Electrodes were placed on the scalp of his patients, which would then transfer electrical information to an EEG (electro-encephalograph) machine. The subjects were told to bend their right index finger. Several events then occurred in succession.

First, the brain readied itself for a specific command from the "will" of the person, who had been told to move his right index finger. About 800 milliseconds before the bending of the finger, the whole cerebral cortex (both left and right lobes) of the brain "lit up" (showed a rise in electrical potential). This was surprising because only the left motor cortex can tell a finger on the right hand to move. Furthermore, this was a long time before the finger was actually moved. Therefore, it was viewed as the time when the intellect or soul of the person was reviewing the situation. The "will" to move the finger and the actual movement itself would occur at a later time. The invisible, non-physical "I" within the person would at a later time be telling the physical brain when, or if, to move the finger.

Second, about 50 milliseconds before the finger actually moved, the EEG machine again showed a sharp rise in electrical activity. This activity however was located only on the left side of the brain, the side that tells the right finger to bend. At this time, "the will" was being exercised to accomplish the bending of the right index finger.

In summary, the sequence of events ran like this. About 800 milliseconds before the movement, the whole motor cortex of the brain lights up to ready itself for a command. Then, about 50 milliseconds before the movement, the mind of the person tells or "wills" the right index finger to move. Finally, at 0 milliseconds, the right index finger actually moves.

(For more on this see: "Attention, Readiness for Action, and the Stages of Voluntary Decision" in *Experimental Brain Research*, Supplement 9, 420-429 [1984].)

Part Four:

Related Topics

32.

Massless Particles and the Soul

Electrons move at a very high rate of speed, in orbits, around the center of each atom in the universe. Electrons can and do jump between orbits, a process that involves energy. If electrons jump to an outer orbital, they require more energy. But if they jump to an inner orbital, they give up energy. This energy is released as a tiny packet of light energy, or what scientists call a "photon." Photons are particles without "mass."

Photons are mass-less particles of light. They are the basis of "light" in our world. We see them all around us. We are awash in light photons. But what does it mean when it is said that photons are without "mass," or are "massless?" The answer is quite mysterious!

The mass of an atomic particle is often confused with its weight. Weight measures the force of gravity acting upon a body. Mass is something different: it is the amount of matter in an object.

Matter is what the physical sciences study. Photons, however, have "no rest mass." In other words, photons have no matter! That is possibly why they can travel "at the speed of light" (186,000 miles per second in a vacuum)!

Photons are "particles" without mass! They exist, but without matter! Notice also that such an expression is exactly what is said about the "soul," namely, that the soul exists, but not in a material way. The soul, like photons, has no matter. It exists *in* matter, but not in a material way!

The science of photons developed rapidly between 1895 and 1915. In 1895, it was thought that electricity was a "fluid" that moved through copper wire. Then, in 1895, J. J. Thomson in England removed all the air from a long glass tube. He then put an electric charge at one end of his vacuum tube. Surprisingly, the electricity traveled through the tube to a positive [+] terminal at the other end! Since there was no medium for an electric "fluid" to travel through, Thomson concluded that only particles carrying an electrical [-] charge could have made the journey. He named these particles "electrons."

In 1905, Albert Einstein of Germany discovered and named the "photon" particle. Einstein learned that "light" hitting certain kinds of metal surfaces caused electrons to be ejected. This was called the photoelectric effect. In Einstein's paper, he concluded that only "particles" could knock electron "particles" out of a metal surface. He called these massless particles "photons." (It is still a mystery how particles without mass can have the momentum necessary to knock electron particles out of a metal substrate.) Einstein received the Nobel Prize for this and other work in 1921.

In 1909, Ernest Rutherford directed his famous "gold foil experiment" in England, which identified the existence of the "nucleus" within atoms. Earlier, in 1899, Rutherford had discovered tiny radioactive particles being ejected from Uranium called "Alpha particles." In 1909, Rutherford's students shot these alpha particles through a very thin layer of gold foil. It was discovered that most of the particles went straight through the foil. Only a few particles were repelled off to the left or to the right, and almost none bounced straight back. Rutherford concluded that the atoms of gold foil (and of all atoms) consisted mostly of empty space!

In 1911, after studying the results of his experiment, Rutherford presented a new model of the atom. He said there was a tiny center in the middle of all atoms with a positive charge. He estimated that the center of an atom was only 1/3,000 the size of one whole atom. He stated further that flying around the nucleus of the atom were electron particles, like planets around the sun. His was known as the "planetary model." Later, it became known that the nucleus of the hydrogen atom was only 1/145,000 the size of the whole atom; while in uranium, the largest atom, it was only about 1/23,000. Therefore, *an atom is almost all empty space*!

Rutherford, however, was not able to explain

why negative [-] electron particles would not spin into the positive [+] charged nucleus. Then, in 1913, Niels Bohr in Denmark offered an improved model of the structure of an atom. In it, the massless photon particles were shown to play a key role. Bohr said that the electron particles traveled in fixed energy paths around the atom. He called these fixed energy levels "shells." In shell #1 there could be only 2 electrons. Shell #2 could hold only 8 electrons. In shell #3 the maximum number would be 18. In shell #4 the number went up to 32. Also, he hypothesized that if an electron decreased its energy level, it would move closer to the center of an atom, say from shell #3 to shell #2. In the process, a "photon," or massless light particle, would be given off.

Then, in 1919, E. Rutherford discovered the "proton" unit within the nucleus of an atom. He did this by firing alpha particles into nitrogen gas (atom # 7), splitting the nitrogen atoms, and creating some new hydrogen atoms (atom #1). This suggested to Rutherford that the hydrogen atom was the fundamental basis of all other atoms, and so he named the molecule at its center a "proton" (meaning "first"). It turns out that all atoms have protons in their nucleus. Atom #6, carbon, has 6 protons. Atom #26, iron, has 26 protons.

In 1932, James Chadwick, a student of Rutherford, discovered the neutron, the second component of the atomic nucleus. All atoms, save hydrogen, have neutrons. Neutrons were difficult to find because they have no electric charge.

In 1964, Murry Gell-Mann at Cal Tech near Los

Angeles predicted the existence of "quark" particles within protons and neutrons. His mathematical work was confirmed in 1968 using the Stanford Linear Accelerator in Northern California. It was then speculated further that the quarks would be held together by a kind of glue: the "gluons."

Gluons are massless particles that were finally found to exist in 1979 using the Petra Collider in Hamburg, Germany. Gluons hold 3 quarks together inside every proton and neutron particle. They mediate the "strong nuclear force" within the nucleus of all atoms. This force is 137 times stronger than the electromagnetic force mediated by the massless photon particles in the outer reaches of an atom's structure.

Thus, two massless particles, the photon and the gluon, are found within every atom in the universe and in every atom of a person's body. Such a fact should cause thoughtful individuals to think about the mystery of one's own self, including one's own soul. Massless particles in every atom! This must surely make one think of the massless nature of the human soul. The immaterial soul is not that much different, in principal, from the immaterial quality of the photon or "light" particle or the gluon particle within an atom's nucleus.

Two thousand years ago, Christ made some amazing statements about "light." First, he said that he was the real light of the universe, the divine reality behind all light in nature. Then he said that humans, like himself, actually have this light of eternal life. They have it, in other words, in their very souls – being created at conception by God himself. Here are his words.

> "I [Jesus] am the light of the world."
>
> John 8:12 & 9:5

> "You are the light of the world."
>
> Matthew 5:14

> "Children of God ... shine as lights in the world ..."
>
> Philippians 2:15

33.

Different Bodies During Life

We are told by science that a person recycles his or her body about once every seven or so years. So, a human being has about twelve different bodies during his or her lifetime! Here is a list of the different bodies that one person can expect to have.

1. Fetal body inside of one's mother
2. Baby body during year 1
3. Toddler body during years 2-3
4. Pre-school body during years 4-5
5. Elementary school body during years 6-12
6. Teenage body during years 13-19
7. Young adult body during years 20-29
8. Mature adult body of age 30-39
9. The forties body of years 40-49
10. Middle age body of 50-59

11. Aging body of 60-69
12. Old body of 70-79
13. Very old body of age 80-89
14. Extremely old body of 90-100

With so many bodies, it is proper to ask: What holds them together? For some, the one answer to this question is the faculty of "memory." Our many unique memories are what constitutes us as one person. Memories help pull our past together so that we can think of our self as one person. And, certainly, this is an important factor. How important it is can be seen when a person is hit in the head and loses all memory. Amnesia is a considerable problem. In older people, a similar situation is seen in Alzheimer's disease, where a patient may not recognize his or her own children. So, the many areas in the brain where memories are stored are important.

However, there is another side to this topic of memories. Memory may not be as extensive a faculty as is sometimes argued. It may also not be as reliable as once thought!

Let's do a little test. Pick one specific year in your life. Now, as best you can, try to remember everything that happened to you during that year. There were 365 days in the year you selected. What, or how much, do you really remember from that year in your life? If you are like me, not very much! And do this two or three more times, picking at one point only one day to remember. What I find when I do this test is that I remember only a few things that happened each year, while most happenings I have forgotten. So, memory may not be primarily what holds us together!

The problem with memory may be related to our opening point: that our body is recycled every seven years. The brain is part of the body, so one can see how memories could possibly be lost.

This whole topic of memory calls to mind a comment made by a young man about his grandfather, who had Alzheimer's disease. The older man could not even remember his grandson's name. Of this fact, the younger man commented: "But he was still in there!" He was the same person! Transcending personality or even memory is the core of our very being, the ineffable essence that is our person.

There is something besides the body and its memory, internal to the body, and to a large extent independent of it, leading a life of its own. That something or someone is the soul. Many people believe that it is the "soul," not primarily the memory, that makes a person "one." Whatever body we may be in at any point in time, whether from number 1 to number 14, the soul mediates that oneness.

The soul, however, is not to be viewed as a separate entity. Christians see the soul as the "form" of the body (see the Council of Vienna, 1312 AD). The soul is not separate from the body; rather, the situation resembles that of two sides of one coin. The brain and the soul work as one together in the service of the person. The brain will die when its parts wear out and stop working properly. The soul, however, which has no parts, will live on. The record of the brain will, nonetheless, somehow be preserved. How exactly this is done, in an immaterial soul, is not known.

One final point: Those individuals who have experienced death, and then have been revived, retain their memory into the next world. (Such "death and return" experiences will be discussed in a following chapter.) What's important here is that memory is not lost in death, when the person leaves his or her physical body. In fact, many of these people experienced a "review" phase when out of their body, which was something like a motion picture review of their entire life. In these reviews, more details appeared to them than they had been able to remember when on earth!

34.

The Split Brain

In the 1960s, Dr. Roger Sperry at Cal Tech (the California Institute of Technology) near Los Angeles studied patients who had undergone an extreme form of surgery to stop epileptic seizures. In this form of surgery, the nerve trunk that connects the right and left lobes (or hemispheres) of the brain is cut. This nerve trunk is called the "Corpus Callosum." Every human being has this structure which allows the brain to work harmoniously as one unit.

But cutting this structure created an abnormal situation. Dr. Sperry found that if he taught something to one hemisphere of the brain, that side would learn it, but the other hemisphere of the brain would know nothing of what had been taught! In other words, Dr. Sperry found that the left brain of these patients does not always know exactly what the right brain is perceiving or has learned (and vice versa). Summarizing, he found that, if the Corpus Callosum had been cut, he could teach one side of the brain something that the other side of the brain was not aware of.

In the human body, the right brain usually controls the left side of the body, and the left brain controls the right side of the body. Thus, theoretically, in these Corpus Callosum cases, one's left arm could be trying to take off a shirt, while the right arm is trying to put it back on! However, even in such extreme circumstances, with the Corpus Callosum cut, this almost never happens. In fact, what usually happens is that one hemisphere of the severed brain overrides the other hemisphere. Thus, even in such extreme circumstances, the brain functions as one entity.

Some, however, using Dr. Sperry's data, have controversially stated that such split-brain patients really have two conscious experiences rather than only one. This, of course, is not true. These patients remain essentially whole or one in their sense of who they are and what they are doing. They have only one consciousness, though divided it may be in small ways, as shown in specific situations in Dr. Sperry's studies.

It is also true that while many investigators believe that while each hemisphere of the brain has separate functions, these are normally integrated by the brain itself. In other words, the brain usually brings together its abilities rather than separating them. Dr. J. Levy of the University of Chicago believes even further that no human activity ever uses just one hemisphere of the brain.

Thus, the brain, like the soul, is essentially one.

It may even be that the soul pushes the brain in this direction of unanimity.

35.

Terminal Lucidity

When a person dies, his or her soul leaves the physical body. The physical body may have been compromised by a serious brain disorder, such as Alzheimer's disease, dementia, a brain tumor, or stroke. The brain tissues affected by these diseases may have been literally destroyed, or are in very poor condition, at death.

When the soul has departed its earthly body, its new situation will be that of a spiritual body, without disease or pain. It will be as if the one who has died is a new person!

> It [the body] is sown in dishonor ...
> it is raised in power; it is sown a soul-
> like body, it is raised a spiritual body.
>
> 1 Corinthians 15:43-44

> God himself will be with them;
> he will wipe every tear from their eyes.
> Death will be no more; mourning

and crying and pain will be no more ...

Revelation 20:3-4

In other words, the person who was very ill before death is totally recovered, or is "raised up," so to speak.

"Terminal Lucidity" is a term that describes a surprising event that mimics this "raised up" condition of one's new life after death. The dying person, for a time, regains mental clarity together with an acute memory while remaining in a diseased or dying body. The Alzheimer researcher Dr. Rudolph Tanzi of Harvard University described this surprising occurrence in the following words.

> Alzheimer's patients who were barely conscious, barely responsive ... suddenly ...just before death say their goodbyes to loved ones, remembering their names ... recalling an event after a decade [of not remembering] ... having first lost their short term memory and then their long term memory. It is a complete mystery! But it is undeniable that it happens, and it is amazing.

"Exploring Frontiers of Biology —
Michael Nahm's Website"
(http://www.michaelnahm.com/terminal-lucidity)
Accessed March 28, 2018

What follows are three examples from documented cases as reported by the biologist, Dr. Michael Nahm.

First, a 91-year-old woman had suffered from Alzheimer's disease for fifteen years. This woman had not recognized her own daughter, or anyone else, for the last five years. However, a few hours before her death she began a very normal conversation with her daughter. She spoke of her fear of death, and the problems she had experienced earlier with the church, and with certain of her family members. Then she lay back down in bed and died.

Second is the case of a young man who had acquired lung cancer that had later spread to his brain. A brain scan near the end of his life showed that this cancer had destroyed most of his brain tissue. The tumor now occupied most of the space where the brain had been. In the days before his death, he was unable to speak, or even to move in bed. However, an hour before he died, he woke up and said goodbye to his family, speaking with them for about five minutes, while patting their hands. This occurrence was reported by the attending nurse and the patient's wife.

Third is the case of an old woman who had suffered from two strokes. The first stroke had paralyzed the left side of her body, and had deprived her of the ability to speak clearly. Then, after a few months, a second

> stroke paralyzed her entire body and left her totally without the ability to speak. However, before this woman died, she sat up in bed without any apparent effort, her face shining brightly (even though her facial muscles had been frozen since the second stroke), and then raised her arms and called out the name of her husband in a clear, joyous tone. After this, she lay back down in bed, and died.

These occurrences are a little experience of Heaven, while still on earth. They demonstrate that people are a composite of body and soul, and while the body may be dying, the soul is alive and well. The soul momentarily escapes a dying body and announces itself!

36.

Near-Death Experiences

The term "near-death experience" was first used by Dr. Raymond Moody in his seminal book, *Life After Life*, which was published in 1975. The term NDE refers to the death of specific individuals who go on to experience life out of their earthly body for tens of minutes or more before being successfully resuscitated and returned to life here on this planet. Because individual persons can be resuscitated, the world can investigate and learn more about events that transpire beyond death. While each NDE is unique in content, there are elements that are common to most of them. The following steps are present in a typical Near-Death Experience.

1. Hearing oneself pronounced dead. This involves a lack of heartbeat, indicated by a flat line on an electrocardiograph recorder. It also involves a non-functional brain that has shut down, as indicated by the flat line on

an electroencephalograph. Furthermore, the eyes appear glassy, the body is cold and grey, and all breathing has stopped.

2. The person then discovers that he or she is outside of their physical body, in a new but identical spiritual body, floating or moving above the scene of their death. Later, the suspended person will be able to provide many details of the actions below of those present, who were working on his or her resuscitation.
3. The person is immediately surprised by overwhelming feelings of peace and quiet. And all pain is gone.
4. Next there is a movement into and through a dark tunnel or space. According to one NDE researcher, about two percent of those entering the dark space are "unable to escape." This is known as the "hell experience" (Lommel). The exact number of those who experience this is unknown because of the shame or guilt associated with it. The return of such individuals may be due to the need for correction within their lives.

5. After passing through darkness or a dark space, the person then begins to meet other spiritual beings — relatives or loved ones who have passed on before. In this situation there is a significant upwelling of love, increased even further by the next step.
6. Then, amidst these spiritual beings, the person begins to see and move toward a bright source of light. This light eventually becomes the "being of light," identified by many as God or Jesus ("I am the light of the world," John 8:12). In this light is also experienced an overwhelming feeling of love. All people who have experienced a NDE speak of it. This love emanates from the "Being of light" for, as is revealed in the Bible, God is both light and love
7. At this point, a "panoramic life review" takes place. The person's whole life from birth to death is seen, like a fast movie. Both good and bad events appear in this review. Directed by this information, some persons are advised to go back to earth to make corrections or to complete their life's work.

8. Lastly, a “border” is reached. If the person goes beyond this point, he or she will not be able to return to earth. What lies beyond this point is not revealed. Those with little children yet to be raised are often advised by deceased relatives or other spirit beings to return to the earth.
9. The NDE ends. The person is resuscitated and returns to their body on earth

How the community on earth views these NDEs varies. What follows are a few examples of historic cases and some unanswerable happenings.

[1] In his book, *Consciousness Beyond Death* (2010), the cardiologist Pim Van Lommel from the Netherlands describes a woman born blind named Vicke. (Vicke’s NDE was also featured in a BBC documentary, “The Day I Died.”)

In 1973, the 22-year-old blind woman was thrown out of a car in a traffic accident. She received a skull fracture, a broken neck, and was in a coma. Her brain had shut down. Totally blind before the accident, she presently found herself above the hospital staff viewing her body which lay on a metal gurney. She now had excellent vision. She tried to talk with the resuscitation staff, but no one could see or hear her. She then migrated through the ceiling “as though it were nothing.” She eventually encountered two blind school mates who had died years before, at the ages of 6 and 11. They had been retarded and blind, but were now bright, healthy, and also able to see!

[2] In his book, *Life After Life* (1975), Dr. Raymond Moody discussed a girl named "Kathy." While Kathy was lying dead in one part of the hospital, she migrated out of her physical body and went into another area of the hospital where her older sister had been crying and was heard to say: "Kathy, please don't die, please don't die." Later, after Kathy had been resuscitated and had returned from her spiritual body, the older sister was shocked to hear Kathy tell her "exactly where she had been (in the hospital) and what she had been saying in her tears."

[3] Dr. Elizabeth Kubler-Ross, a psychiatrist from Switzerland who later taught at the University of Chicago, penned a book in 1991 entitled, *On Life After Death*. In this book, Kubler-Ross speaks of a 12-year-old girl who underwent a Near-Death Experience. After the event was over, the girl spoke of the experience with her father. She described an extremely beautiful landscape but then said that one thing especially bothered her. She said that she had met and was "comforted by her brother" in Heaven. She, however, was unaware that she had a brother! Her father then told her, that, in fact, she did have a brother, who had died three months before she was born! The parents had not told the girl of this event,

probably to shield her from its sadness.

[4] In 2014, a book by Judy Bachrach entitled *Glimpsing Heaven* was published by the National Geographic Society. In this book, the author tells the story of the NDE of Pamela Reynolds, who underwent a long operation for a ballooning artery at the base of her brain. During the operation, Ms. Reynolds's body was put into an inactive state by being cooled to 60 degrees F. (40 degrees below normal). She was also made to be unconscious due to an extensive use of anesthesia ("a barbiturate coma"). And the brain was straight-lined on a brain-wave machine, while a heart-lung machine pumped her blood. Lastly, her eyes were taped shut and plugs were placed in her ears!

Pamela Reynolds was nonetheless completely conscious during and after her surgery while being in a "spiritual body." After she regained her physical ability to talk, she told her surgeon, Dr. Robert Spetzler, many facts about the operation. She told him how the artery in her right leg had been too small to be hooked up to the heart-lung machine, so that they had to use an artery in her left leg. She accurately described the saw that was used to cut into her scalp, describing it as looking like an electric toothbrush. She said further that her doctors had listened to a tune called "Hotel California" by the Eagles during her surgery, and that its lyrics had stated: "You can check out anytime you like!" And she said that at the very end of the operation her heart had to be shocked twice to restart it.

She also described how, during the mid-portion of the surgery, she had moved up and out of the operating theater. She had been drawn toward a "shower of light," and then met several deceased relatives on the journey, one of them being her beloved uncle, Gene Saxon, who had taught her how to play the guitar and had died at age 39. Another was her grandmother, "Marie," who, when asked if the light was God, said: "Oh no, Baby! ... The light is what happens when God breathes!" These two individuals eventually persuaded Pamela Reynolds to return to her earthly body to care for her young children.

[5] Another account of an NDE found in Ms. Bachrach's book is that of Dr. Anthony Cecoria, a New York orthopedic surgeon. In August 1994, Dr. Cecoria, then 42, was attending an outdoor family reunion near Albany, New York. Dr. Cecoria had walked to a phone booth near the park pavilion to call his mother. While using the pay phone, lightning struck the phone booth and hit Dr. Cecoria on the left side of his lips. The lightning then traveled down his left side and out his left foot. He was immediately knocked down unconscious. While unconscious, he saw his mother-in-law run for help, and about the same time felt himself "go forward" propelled by some unseen force. Once above his physical body, he looked down at himself on the ground, and thought: "I'm dead!" He then moved up the stairs to the pavilion and went through its walls into the interior of the building, where he saw his children, ages 4, 5,

and 6, having their faces painted. He thought: "I'm never going to see them again." At this point, he began to float out of the room toward a "bluish white light."

Later, after Dr. Cecoria had been successfully resuscitated and was in the hospital, he asked his wife the following question: Were the kids in that pavilion building having their faces painted when this happened? She answered, "Yes"!

[6] Dr. Pim Van Lommel in his book *Consciousness Beyond Life* (2010) relates the following account of an out-of-body experience of a 44-year-old comatose man. This account also appears in the British science journal, "Lancet" 358 (2001).

The 44-year-old man had been found in a park by passers-by and had received defibrillation (re-activating the heart by electric shock) and was put on a machine to breathe for him. While doing the latter procedure, the attending nurse had to remove the upper set of dentures (false teeth) from the comatose man's mouth. Then, after doing this, she put the dentures on a nearby cart.

The patient remained in a coma for over a week. Then, after regaining consciousness, he was put in a coronary care unit in the hospital. When this same nurse came by to give the patient his medication, he said to her: "You know where my dentures are!" The nurse was very surprised. Then, he told her: "You were there when they brought me into the hospital, and you took the dentures out of my mouth and put them on the cart; it had all those bottles on it, and there was a sliding drawer underneath, and you put my teeth there." The nurse said in the Lancet article: "I was all the more amazed because I remembered this happening when the man was in a deep coma and undergoing resuscitation." As it turned out, the patient had been watching his own resuscitation from above and was able to give a "detailed description of the small room" he was in. He had also heard his doctors discuss the meager possibility that he would survive.

[7] The last example of a NDE will be left for the reader to investigate. It is in the book *Proof of Heaven: A Neurosurgeon's Journey into the Afterlife* (2012) by Dr. Eben Alexander. Dr. Alexander had taught at Harvard University for 15 years before he had his NDE. His journey into the afterlife took place during the week he lay in a coma, due to a rare E. Coli infection of his brain. Dr. Raymond Moody, author of "Life After Life," said of this book: "Dr. Eben Alexander's near-death experience is the most astounding I have heard in more than four decades of studying this phenomenon."

The individual death experiences narrated within this chapter cannot be dismissed as mere hallucinations, since no one can hallucinate without a functioning brain.

> There are two ways to be fooled. One is to believe what isn't true; the other is to refuse

to believe what is true.

Soren Kierkegard (1813-1855)

37.

The Soul's Transcendent Reach

The individual soul is the work of God. Since the soul is immaterial, it cannot have had a material cause for its existence. The creation of the soul, therefore, is the proper work of God. Each soul is brought into existence solely by God (Genesis 2:7).

The soul of a person has certain characteristics that demonstrate its origin from God. These are the so-called "transcendentals" that refer to the soul's upward movement toward perfect truth, perfect love, perfect goodness, perfect beauty, and a perfect home.

The Greek thinker, Plato, mentioned these transcendental powers in his dialogue, *Phaedrus*, published in 370 B.C., saying:

> ... the ability of the soul [is] to soar up to heaven to behold beauty, truth, goodness and the like. [24]

These transcendental powers could not have come solely from within the human person. The cause of such unearth-like activity must have come from God – a top-down occurrence. In this chapter, four of these transcendentals will be discussed, as they reach upward into Heaven and point to God.

[1] Perfect Goodness

Human beings know what good and evil are. They also have a "sense" of what perfect goodness and perfect justice should be. This sense of perfect goodness and justice is, however, often separate from what is actually present here on earth. What should be done is often known, even though it has never before been done. In the quiet of one's soul, or conscience, is seen what does not yet exist. It is this "seeing" of the good that doesn't yet exist that identifies the presence of our transcendent soul.

This strong movement toward perfect goodness and justice is present early in life. When a child sees a parent do something unjust, they will immediately say: "That's not fair!" They seem to immediately know what "fair" really means. But where, at such an early age, does this knowledge come from? It has not been learned from a book or at school!

Adults behave in similar ways. Adults possess a "not of this world" sense of what ought to be happening in situations. They are sometimes very surprised that others do not see the "obvious." They may even become shocked or outraged at the deficit in goodness or justice around them.

This striving for the "perfect" answer to a variety of problems in the world is a good thing. Nonetheless, the reality is that our expectations for a perfect environment, or social order, or legal system, cannot be achieved in this finite and fallen world that we live in. Yet, because of the powers of our spiritual soul, we keep on trying to reach the perfect. We have to! Because our soul keeps on telling us what is tending toward perfection, and what should therefore be put in place. Our desire is guided by a transcendent awareness of perfect goodness and perfect justice. Human beings have this self-transcendent idea of "the perfect" because God has given a soul to every person at the moment of their conception.

[2] Perfect Truth

Truth cannot be cornered or arrested. Truth remains elusive. The search for truth begins early in life. Young children ask endless questions, and the answers given them never quite satisfy them. There is always more to know, more questions to ask!

This pattern remains true throughout life. The answers to every topic in this world remain incomplete; even when the most comprehensive answers are given, there is still more to know. There is always more work to do in the endless quest to complete a truth. This ability to know that things are not at their highest level of possible truth is an otherworldly trait. It is not a natural, but rather a supernatural kind of insight. It is an example of the power of the soul to know things even if they are not in this world. The soul is an otherworldly instrument. The worldly answer does not completely satisfy it.

[3] Perfect Beauty

Humans have a tendency to idolize beauty; that is, they turn the beautiful into a kind of idol, and then try to worship it. Some people would, for example, prefer to go to an art museum, rather than to a church, to worship God. Another example of this is seen when Hollywood actors are treated as "stars," to be viewed in a divine way, like the stars in the night sky.

There is, however, a problem. Art, in the long run, does not completely satisfy. Humans keep on trying to make artistic expressions perfect — expressions that will approach the perfections of God. This, of course, does not happen. Nonetheless, people continue to have the "perfect" idea of art, if only it could be achieved here on earth! When a musical composition or a movie does in fact approach perfection, some still believe that more could have been done to make it even better.

This problem with beauty is not just about "great works of art." The problem is seen everywhere. We do not look good enough in the mirror, and others around us believe that they do not look good enough, either. It is informative to see pictures of a woman before she puts on her makeup and after. One can look at jewelry and see that there are endless ways to design and craft beautiful pieces. The human soul keeps on reaching out for the perfect, coming up with totally new expressions of beauty in clothes, in building design, musical composition, performances, novels, plays, and drawings – all having not existed previously in this world.

Occasionally, a person will discover a novel, painting, musical composition, mountain or sunset that is truly beautiful in a way that approaches perfection. Nonetheless, over time, that person will get bored with it and will try to put something even more attractive in its place.

Human beings have an ability to imagine what is even more beautiful than the beauty that is before them. They are soulfully aware that a sunset might be a little more beautiful, or that a national park might be *even better*. This built-in creative sense creates in people the desire for an even more perfect experience of beauty.

All manifestations of beauty here on earth are imperfect. This fact is not solely learned here on earth. Rather, it is an "insight" directly due to the power of one's soul. The soul is a heavenly instrument that senses and knows beauty from a heavenly perspective. It is not, therefore, surprising that the greatest examples of beauty and the greatest works of art summon up feelings of the divine in us. This sense of perfect beauty in the human consciousness reveals a spiritual, self-transcendent, or soul-presence within all human persons.

[4] Perfect Home

Probably no human being has gone through life without experiencing some feelings of separation from other people and even from the world itself. At various moments in life, we have probably felt a certain aloneness and isolation. At times, the world around us does not seem to be our real home. An extreme form of this feeling was expressed by T.S. Eliot in his play, *The Cocktail Party*:

> [What] has happened to me has made me aware
> That I have always been alone. That one always is alone....
> Everyone's alone—or so it seems to me.
> They make noises, and think they are talking to each other;
> They make faces, and think they understand each other.
> And I am sure that they don't....

A less extreme, more generalized expression of this idea, was offered by another English writer, C.S. Lewis.:

> I have found a desire within myself that no experience in this world can satisfy; therefore the most probable explanation is that I was made for another world.

Where does this "desire within myself," of which

Lewis speaks, come from? Where does this sense of "not belonging" here on earth come from? The answer, of course, is not from "the earth." Even rich persons on this planet, who can buy every amenity this world offers, can and do experience what Lewis is speaking of. Its source, therefore, is something deeper. The Bible says that it emanates from God, who has prepared a better home in Heaven for every human being.

> In my Father's house
> there are many dwelling places ...
> I [Jesus] go to prepare a place for you.
> John 14:2-3

Our soul, the internal "form" of us, hears this voice of the Lord. The soul can access this supernatural information which the world cannot provide. C.S. Lewis has left us another quotation which explains why our soul reaches out to God to address this problem of aloneness.

> God cannot give us a happiness and peace apart from Himself ... Aim at Heaven and you will get earth thrown in.

Conclusion

The "transcendentals" discussed above "transcend" what is normal in human life. They represent movements toward the realm of the perfect, which is God. They speak of upward movements toward perfect goodness, perfect truth, perfect beauty, and a perfect home. They reach upward and beyond what is possible in a natural way in this world. They take one into the supernatural or spiritual realm that the soul is capable of penetrating. They are strong evidences for the existence and power of the human soul.

38.

Love's Journey Home

Love is another elusive reality here on earth. Love is not primarily an emotion or a feeling. Love is related to a person's orientation in life. How a person "does life" will determine how a person will love and be loved.

There are four principal life orientations. 1. The materialistic, or "pleasure oriented" life style; 2. The competitive, inward, or "I centered" (ego-in) life style; 3. The contributive, or outward, "other centered" (ego-out) life style; and 4. The transcendent, spiritual, or "upward oriented" life style, revealing the human movement toward Heaven. Each of these life orientations will determine the kind of love that will be present in one's life.

Levels 1 and 2 are about the selfish person. This person's activities are self-beneficial (in other words, "selfish). This is the "taker" in everyday life. In this lifestyle, pleasure, utilitarian friendships, and romantic love will be ends in themselves.

By contrast, the Level 3 person reaches out to the

world, tries to make a contribution in other people's lives, and wants to make the world a better place. This is the "giver" in everyday life. Here the command to "love your neighbor as yourself" becomes operative. Love becomes outgoing, and helpful; it is about others as well as oneself. Friendships and romantic love are no longer ends that mostly benefit "me."

Level 4 is an extension of Level 3. Level 4 is transcendent in nature. The Level 4 life orientation reaches upward, and outward toward eternity. The overriding kind of love here is the "love of God." It is a selfless, sacrificial, an unconditional kind of love ("agape"). This kind of life orientation moves upward toward an eternal, painless experience in Heaven.

What follows is a look at levels 1-4, to identify some of the characteristic behaviors of each level.

Level 1: Love of Self

A person's interest here is to acquire many material possessions (bank accounts, an expensive house and car, fine food and clothes, and so on), the need to be entertained (movies, sports events, snowboarding, travel), and access to sensual pleasures. To this kind of person, "freedom" means having cash and other things, when they want them. Affection is good, especially when they do not have to help anyone. There is no interest in sacrificing.

Friendships for someone with this life orientation do not arise from empathy, or a desire to do good for someone else. Friendships are utilitarian. This person has little or no interest in the "common good" (of Level 3) or in the ultimate good (of Level 4). Suffering is viewed as a negative, because it deprives one of pleasure. If they suffer, such a person will try to erase the feeling by eating, drinking alcohol, having sex, or shopping. They also have little interest in ethics.

Level 2: Love within the Competitive World

This person seeks a high quality of life and is very competitive. There is a push to win at everything: self-esteem, status, power, and control. Talents are developed through education with the goal to stand out and to be respected. Such a person seeks a high prestige career, high academic achievements, influence in public, and power in the family. He or she feels freest when winning over others. Such people seek to achieve goals to benefit themselves. They do not feel free when helping others and do not want to spend time with those who cannot help advance their own goals. Such a person embraces stoic virtues of self-discipline, courage, and perseverance for self-centered ends (not to help others). They want to feel superior to others, and are not much interested in conscience, or justice. They have a narcissistic need to be loved but have little interest in sharing with another person. Intellectual friends help to reinforce one's own intelligence. Elitist attitudes are common. Suffering is perceived as slipping into mediocrity.

Level 3: Love of Others

This person uses one's talents, energy, and time to make a positive difference in the world. They wish to create a legacy. Their focus shifts from getting what they want (Level 1) or being in control and on top (Level 2), to doing good for someone or something beyond oneself (Level 3). Self-centered stoic virtues (Level 2) are replaced by other-centered virtues involving sacrifice (Level 3). They develop good empathetic relationships where giving is as important as receiving. They wish to advance a noble purpose and have a positive impact on family, friends, community, church, organizations, and culture. They feel free, even when exercising discipline and self-sacrifice. They subordinate individual good to the common good. They are empathetic. Self-sacrificial love (*agape*) exists within their friendships and within their romantic love. They have the ability to love others rather than the need to be loved (narcissism). They believe that good ends do not justify evil means. Nothing justifies hurting others; thus, abortion and euthanasia are rejected. They believe that all humans should be fully developed. They understand that suffering can help remove one from a superficial life, make one more tolerant of the weaknesses in others, and decrease attitudes of superiority.

Level 4: Love of God

Transcendental love moves the "love of neighbor" (Level 3) into the realm of the "love of God" (Level 4). "God is Love" (1 John 4:8), and God's love has an eternal dimension to it that spreads over the sacrificial love or "agape" of Level 3. God's love will redeem all suffering, even death. Humans are the possessions of God, intended to belong in God's family. Level 4 persons acknowledge the basic goodness, lovability, and mystery of all humans because of their being created in God's image (Genesis 1:26). This type of person gives up what they want (Level 1), the goal for status and power (Level 2), and of doing limited good for others (Level 3), in order to answer the call from God to do His work (Level 4). They want not only to help others, but to build up the Kingdom of God, pursue goodness and truth, and thus prepare for Heaven.

What follows are a few quotations that evidence God's basic intention: that all persons should eventually live in Heaven!

God our Savior wishes that all men be
saved and come to a knowledge of the truth.

1 Timothy 2:4

It is not the wish of your Father in Heaven
that one of these little ones be lost.

Matthew 18:14

God sent the Son [Jesus] into the world ...

that the world might be saved through him.

John 3:18

God [the Father] did not intend anger for us,
but [rather] the possession of salvation
through Jesus Christ our Lord.

1 Thessalonians 5:9

The Lord ... is patient, not planning to
destroy some, but that all might change
their mind to make room [for Heaven].

2 Peter 3:9

About the Author

Roger Skrenes studied science as an undergraduate and history as a graduate. He holds a master's degree in religion and has taught high school in Los Angeles, California, for over thirty years, including six summers in the California Youth Authority (a prison for teenage boys in Whittier, CA). He is the father of three adult children, Mary, Mark and Therese.

www.ingramcontent.com/pod-product-compliance
Lightning Source LLC
Chambersburg PA
CBHW022009090426
42741CB00007B/951

* 9 7 8 1 9 5 0 1 0 8 9 3 0 *